Methods of Experimental Physics

VOLUME 8

PROBLEMS AND SOLUTIONS FOR STUDENTS

METHODS OF EXPERIMENTAL PHYSICS:

L. Marton, *Editor-in-Chief*

Claire Marton, *Assistant Editor*

1. Classical Methods, 1959
   Edited by Immanuel Estermann

2. Electronic Methods, 1964
   Edited by E. Bleuler and R. O. Haxby

3. Molecular Physics, 1961
   Edited by Dudley Williams

4. Atomic and Electron Physics—Part A: Atomic Sources and Detectors; Part B: Free Atoms, 1967
   Edited by Vernon W. Hughes and Howard L. Schultz

5. Nuclear Physics (*in two parts*), 1961 and 1963
   Edited by Luke C. L. Yuan and Chien-Shiung Wu

6. Solid State Physics (*in two parts*), 1959
   Edited by K. Lark-Horovitz and Vivian A. Johnson

7. Atomic and Electron Physics—Atomic Interactions (*in two parts*), 1968
   Edited by Benjamin Bederson and Wade L. Fite

8. Problems and Solutions for Students, 1969
   Edited by L. Marton and W. F. Hornyak

Planned volumes:—

Plasma Physics
Edited by H. Griem and R. Lovberg

Volume 8

Problems and Solutions for Students

Edited by

L. MARTON

National Bureau of Standards
Washington, D.C.

and

W. F. HORNYAK

Department of Physics and Astronomy
University of Maryland
College Park, Maryland

1969

ACADEMIC PRESS • New York and London

ACADEMIC PRESS, INC.
111 Fifth Avenue, New York, New York 10003

United Kingdom Edition published by
ACADEMIC PRESS, INC. (LONDON) LTD.
Berkeley Square House, London W1X6BA

LIBRARY OF CONGRESS CATALOG CARD NUMBER: 69-13487

PRINTED IN THE UNITED STATES OF AMERICA

CONTRIBUTORS TO VOLUME 8

Numbers in parentheses indicate the pages on which the authors' contributions begin.

A. BLANC-LAPIERRE, *Directeur du Laboratoire de l'Accélérateur Linéaire, Faculté des Sciences, Paris, France* (165)

H. BRECHNA, *Stanford Linear Accelerator Center, Stanford, California* (55, 61, 63, 65, 67, 77, 81, 93)

H. DANIEL, *Physik-Department der Technischen Hochschule München, Munich, Germany* (187)

W. GENTNER, *Max Planck Institute für Kernphysik, Heidelberg, Germany* (187)

HANS R. GRIEM, *Department of Physics and Astronomy, University of Maryland, College Park, Maryland* (161)

M. J. HIGATSBERGER, *Österreichische Studiengesellschaft für Atom Energie, Vienna, Austria* (169, 177, 181, 191, 203, 213, 227, 233, 239, 245, 249)

W. F. HORNYAK, *Department of Physics and Astronomy, University of Maryland, College Park, Maryland* (11, 23, 253, 267, 271, 277)

I. KAUFMAN, *Electrical Engineering Department, Arizona State University, Tempe, Arizona* (107, 117, 123, 129, 135, 143, 147, 153)

J. F. KOCH, *Department of Physics and Astronomy, University of Maryland, College Park, Maryland* (19)

L. MARTON, *National Bureau of Standards, Washington, D.C.* (163, 165)

J. P. MATHIEU, *Laboratoire des Researches Physiques, Faculté des Sciences, Paris* (39)

M. ROUSSEAU, *Laboratoire des Researches Physiques, Faculté des Sciences, Paris* (39)

ROBERT M. ST. JOHN, *Physics Department, University of Oklahoma, Norman, Oklahoma* (27)

E. A. STERN, *Department of Physics, University of Washington, Seattle, Washington* (23)

D. M. ZIPOY, *Department of Physics and Astronomy, University of Maryland, College Park, Maryland* (1)

FOREWORD

In conversation with the volume editors of our series, the idea came up to add a volume containing problems in experimental physics. Discussions with many physicists confirmed the desirability of providing graduate students in physics with problems oriented toward the feasibility, limits of accuracy, and of precision of experiments, the kind of information an experiment is likely to yield, signal-to-noise ratio and expected signal level of the experiment, the general considerations about the design of an experiment, etc.

The present volume is the result of our endeavor to select relevant problems. We consider it as a first step in the direction indicated above and would like to invite comments and further examples. Hopefully enough, new material will be available in the foreseeable future to justify issuing a revised edition. In the meantime, we wish to express our heartfelt thanks to our contributors and to all those whose kind cooperation and advice helped us to put the volume together.

L. MARTON
W. F. HORNYAK

Washington, D.C.
July 1969

CONTENTS

Contributors to Volume 8 v

Foreword vii

1. The Theoretical Limit for the Detectability of a Signal 1
   D. M. Zipoy

2. Low Frequency Loudspeaker Design 11
   W. F. Hornyak

3. Temperatures below 4.2°K 19
   J. F. Koch

4. Reflectivity Measurement 23
   E. A. Stern and W. F. Hornyak

5. Calibration of a Spectroscopic System 27
   Robert M. St. John

6. Visibility of Young's Interference Fringes 39
   M. Rousseau and J. P. Mathieu

7. Field of Line Currents 55
   H. Brechna

8. Field of a Circular Loop 61
   H. Brechna

9. Dipole Fields 63
   H. Brechna

10. Fields Due to Magnetized Dipoles 65
H. Brechna

11. Magnetic Fields of a Current Loop 67
H. Brechna

12. Pair of Current Loops 77
H. Brechna

13. Field of a Solenoid of Finite Cross-Sectional Area 81
H. Brechna

14. Field of a Pair of Solenoids 93
H. Brechna

15. Broad Band Impedance Matching 107
I. Kaufman

16. Balun and Narrow Band Impedance Matching 117
I. Kaufman

17. Measurement of Pulsed Microwave Power 123
I. Kaufman

18. Microwave Cavity Resonator 129
I. Kaufman

19. Coupling of a Microwave Cavity to a Waveguide 135
I. Kaufman

20. Phase Shifter 143
I. Kaufman

21. Waveguide Interferometer 147
I. Kaufman

22. Microwave Optics 153
I. KAUFMAN

23. Measurement of Gravitational Red Shift 161
HANS R. GRIEM

24. Relativistic Effect on Atomic Clocks 163
L. MARTON

25. Electron Storage Ring 165
L. MARTON AND A. BLANC-LAPIERRE

26. Mass Spectrometer for Upper Atmosphere Research 169
M. J. HIGATSBERGER

27. Mass Spectrometer for Reactor Fuel Research 177
M. J. HIGATSBERGER

28. Isotope Separator 181
M. J. HIGATSBERGER

29. Half-Life of N^{13} 187
H. DANIEL AND W. GENTNER

30. Nuclear Counting Experiments 191
M. J. HIGATSBERGER

31. Gamma Spectroscopy 203
M. J. HIGATSBERGER

32. Spectrometer Comparison 213
M. J. HIGATSBERGER

33. Pair Production Cross Section 227
M. J. HIGATSBERGER

34. Reactivity Oscillator 233
M. J. HIGATSBERGER

35. Neutron Transmission Measurements 239
M. J. HIGATSBERGER

36. Fuel Irradiation Capsule 245
M. J. HIGATSBERGER

37. Reactor Regulation and Control 249
M. J. HIGATSBERGER

38. Isochronous Reaction Surfaces 253
W. F. HORNYAK

39. Neutron Threshold Measurements 267
W. F. HORNYAK

40. Experimental Targetry 271
W. F. HORNYAK

41. Dynamics of Mass Exchange Reactions 277
W. F. HORNYAK

I. THE THEORETICAL LIMIT FOR THE DETECTABILITY OF A SIGNAL*

I.I. Problem

It is never possible to measure anything to arbitrarily high precision. The only exceptions to this are some measurements that involve counting. For instance, if you have three walnuts lying in front of you and if you count them you will come up with the number three (hopefully). However any measurement that involves measuring a continuous variable, such as voltage or position, is necessarily inexact *even in principle*. The reason for this is that any detector and associated equipment is made of material and that material is at a finite temperature. The atoms and electrons that make up the material jostle around in a random manner and because of this they give rise to a signal that is essentially indistinguishable from the signal that you are trying to measure. The ideas here can best be described by an example.

Suppose you want to measure a mechanical motion that you know is oscillating at a single frequency. The question is: What is the smallest vibration amplitude that can in principle be measured?

I.2. Solution

As a model for a detector we note that if you hang a weight on a spring and then move the upper end of the spring up and down, the amplitude of the weight will be much larger than the driving amplitude if you are at the resonant frequency of the mass and spring assembly. It is essentially a mechanical amplifier and serves as a good candidate for a sensitive detector (Fig. 1.1). We shall mention how the amplified motion is measured later. The equation of motion is:

$$m\ddot{x} = -K(x-y) - D\dot{x}, \tag{1.1}$$

where the overdot denotes time differentiation. We have assumed that the damping term, $-D\dot{x}$, depends only on the motion of the mass (such as air resistance). Another term, $-D'\dot{y}$, could be added and would take into

* Problem 1 is by D. M. Zipoy.

account the losses in the spring but it would clutter up the following analysis without adding anything essential to it.

Suppose $y(t) = y_0 e^{i\omega t}$. Let $x(t) = x_0 e^{i\omega t}$, then

$$(-\omega^2 m + i\omega D + K)x_0 = K y_0$$

or

$$x_0 = \frac{K y_0}{K - m\omega^2 + i\omega D}. \tag{1.2}$$

This is a maximum when $\omega^2 = K/m \equiv \omega_0^2$ (for D small).

$$x_{0_{\max}} \approx \frac{m\omega_0^2}{i\omega_0 D} y_0 \equiv -iQy_0, \quad \text{where} \quad Q \equiv \frac{m\omega_0}{D}. \tag{1.3}$$

In order to make x_0 large we want to make Q (the "quality" factor) large, that is, we want a large mass and low damping. Even at this point there are

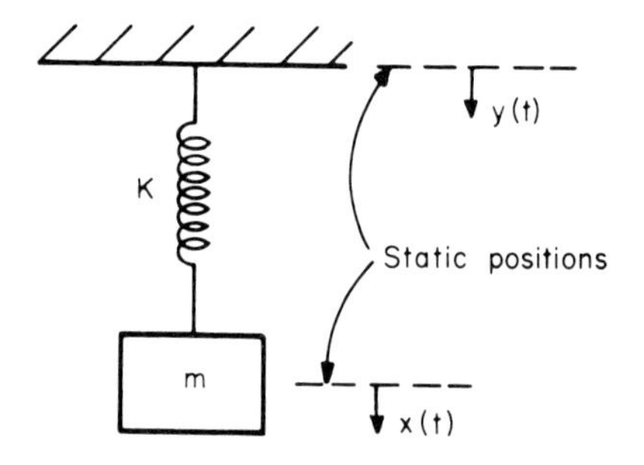

FIG. 1.1. Vibration detector.

limits to how big Q can be. The mass has to be small enough so that it does not cause the support to come crashing down. It is conceivable that the damping D could be made exceedingly small by putting everything in a vacuum and perhaps cooling it to a very low temperature at which point the internal losses in the spring tend to become small. From a practical point of view there is a limit to how large you *want* Q to be. It is determined by your lifetime as an upper limit (we presume). The reason for this limit is the following. At some time or other the detector has to be constructed, hooked onto the support, and released. It takes a while before the oscillation of the mass builds up to its steady state amplitude, x_0; how long is this? You can easily verify that the characteristic exponential buildup time is $2Q/\omega_0$. Now if the period of oscillation is 1 sec and $Q = 10^7$, then the buildup time is about a month and so you would have to wait a few months after you released the mass before you could make a measurement.

Let us now go on to the main point of this problem. What is the thermal noise associated with this device? The term "noise" is used throughout science and denotes *any* random (unpredictable) process or motion. The source of noise arises from the equipartition theorem of statistical mechanics, that innocuous theorem you heard in your elementary physics course and promptly forgot. It is used as follows. You know that each atom in the mass, spring and support have $\frac{1}{2}kT$ of energy in each degree of freedom. That is, they are twitching around such that they have $\frac{1}{2}kT$ of energy in the "x component" of kinetic energy, etc. This concept can be generalized to say that each normal mode of a system constitutes a degree of freedom (actually two degrees of freedom, one for the kinetic and one for the potential energy of the mode of vibration). Remember that there are two ways (at least) of describing the motion of a blob of material; either in terms of the position of each atom as a function of time or by giving the amplitudes and phases of all the possible normal modes of vibration. This latter description is most useful for our purposes because it enables us to immediately write down the mean square fluctuation in position of the mass m. (The lowest mode of this system is just what is usually thought of as *the* resonant mode; there are other modes which involve transverse and longitudinal vibrations of the spring, for instance. In all there are $6N$ modes of the system where N is the number of atoms in the mass, spring, and support!) We will only be concerned with the usual mode since the others are usually at a different enough frequency that they can be discriminated against. We have

$$\tfrac{1}{2}K\langle x^2\rangle_N = \tfrac{1}{2}kT \qquad \text{or} \qquad \langle x^2\rangle_N = \frac{kT}{m\omega_0{}^2}. \tag{1.4}$$

The signal to noise ratio is:

$$\left(\frac{S}{N}\right)^2 \equiv \frac{\frac{1}{2}|x_0|^2}{\langle x^2\rangle_N} = \frac{m\omega_0{}^2Q^2}{2kT}\,y_0{}^2. \tag{1.5}$$

For a numerical example, suppose we are interested in a frequency of 1 Hz. Let $m = 10$ g, $\omega_0 = 2\pi$, $Q = 100$ (a reasonable value), $T = 300°$K, then $S/N \approx 10^{10}\,y_0$. So with the above detector we could measure vibration amplitudes of about 10^{-10} cm. This could be improved by increasing m and Q and decreasing the temperature.

We can now ask the question: What is the frequency distribution of the thermal noise motion of the mass? Is it all at ω_0 or is it spread around over all frequencies? For instance, if we found that it was spread around we might be able to think up a way to only measure the motion at ω and thus eliminate most of the noise.

To answer the above question we make the reasonable assumptions that the noise fluctuations can be considered to be caused by a noise force F_N acting on the mass and that the frequency distribution of this noise force is reasonably smooth. Then:

$$m\ddot{x} + D\dot{x} + Kx = F_N(t). \tag{1.6}$$

Fourier transform the equation. Let

$$x(t) = \frac{1}{(2\pi)^{1/2}} \int_{-\infty}^{\infty} x(\omega)\, e^{i\omega t}\, d\omega, \quad \text{etc.} \tag{1.7}$$

Then,

$$(-m\omega^2 + i\omega D + K)x_N(\omega) = F_N(\omega) \tag{1.8}$$

or

$$x_N(\omega) = \frac{F_N(\omega)}{m} \frac{1}{\omega_0{}^2 - \omega^2 + (i\omega\omega_0/Q)} \tag{1.9}$$

which is peaked near $\omega = \omega_0$. The above hope is not realized. We will complete the calculation anyway:

$$x_N{}^2(t) = \frac{1}{2\pi} \int_{-\infty}^{\infty} \int_{-\infty}^{\infty} x(\omega)x(\omega')e^{i\omega t}\, e^{i\omega' t}\, d\omega\, d\omega' \tag{1.10}$$

$$\langle x_N{}^2(t) \rangle = \frac{kT}{m\omega_0{}^2}$$

$$= \lim_{T\to\infty} \frac{1}{T} \int_{-T/2}^{T/2} x_N{}^2(t)\, dt$$

$$= \lim \frac{1}{2\pi T} \int_{-\infty}^{\infty} d\omega \int_{-\infty}^{\infty} d\omega' x_N(\omega)x_N(\omega') \int_{-T/2}^{T/2} e^{i(\omega+\omega')t}\, dt$$

$$\approx \lim \frac{1}{T} \int_{-\infty}^{\infty} d\omega \int_{-\infty}^{\infty} d\omega' x_N(\omega)x_N(\omega')\, \delta(\omega + \omega')$$

$$= \lim \frac{1}{T} \int_{-\infty}^{\infty} x_N(\omega)x_N(-\omega)\, d\omega$$

$$= \lim \frac{1}{T} \int_{-\infty}^{\infty} |\, x_N(\omega)\,|^2\, d\omega. \tag{1.11}$$

We have used the fact that

$$\lim_{T\to\infty} \frac{1}{2\pi} \int_{-T/2}^{T/2} e^{i\alpha t}\, dt = \delta(\alpha)$$

where $\delta(\alpha)$ is the Dirac δ function. We also used the fact that if $x(t)$ is real then $x(-\omega) = x^*(\omega)$ where the asterisk (*) denotes the complex conjugate. This also implies that $|x(\omega)|^2$ is an even function of ω. Using this fact we can put the integral in a more standard form

$$\langle x_N^2(t) \rangle = \lim_{T\to\infty} \frac{4\pi}{T} \int_0^\infty |x_N(\omega)|^2\, d\nu \tag{1.12}$$

where $\omega = 2\pi\nu$ and ν is the ordinary frequency. The quantity $\lim_{T\to\infty} (4\pi/T)\, |x_N(\omega)|^2$ is called the power spectrum of the displacement $x_N(t)$; we will see that this is finite. We want to evaluate the power spectrum in terms of the temperature, etc. From (1.11) and (1.9),

$$\frac{kT}{m\omega_0^2} = \lim_{T\to\infty} \frac{1}{T} \int_{-\infty}^{\infty} \frac{|F_N(\omega)|^2}{m^2} \frac{d\omega}{(\omega_0^2 - \omega^2)^2 + (\omega\omega_0/Q)^2}\,. \tag{1.13}$$

We see that the integrand is sharply peaked around $\omega = \pm\,\omega_0$ so if we assume that $|F_N(\omega)|^2$ does not vary much over the width of the peak, it can be taken outside of the integral and then the integral can be evaluated most conveniently by the method of residues

$$\frac{kT}{m\omega_0^2} = \lim_{T\to\infty} \frac{1}{T} \frac{|F_N(\omega_0)|^2}{m^2} \int_{-\infty}^{\infty} \frac{d\omega}{(\omega_0^2 - \omega^2)^2 + (\omega\omega_0/Q)^2}$$

$$= \lim_{T\to\infty} \frac{1}{T} \frac{|F_N(\omega_0)|^2}{m^2} \frac{\pi Q}{\omega_0^3}\,. \tag{1.14}$$

Therefore:

$$\lim_{T\to\infty} \frac{4\pi}{T} |F_N(\omega_0)|^2 = 4kTD\,. \tag{1.15}$$

This is the power spectrum of the noise force. We see that ω_0 does not appear explicitly on the righthand side of (1.15) and therefore the noise force can be represented by a flat spectral (frequency) distribution (a so-called "white-noise" spectrum). By analogy with (1.12), the mean square noise force in a small band of frequencies $\Delta\nu$ is

$$\langle F_N{}^2 \rangle_{\Delta\nu} = \int_{\Delta\nu} 4kTD\, d\nu = 4kTD\, \Delta\nu\,. \tag{1.16}$$

This result is very general and can be shown to be exact for low frequencies ($h\nu \ll kT$ where h is Planck's constant). For the sake of completeness we can write down the power spectrum of the displacement $x_N(t)$ from (1.9) and (1.12) .

$$\lim_{T\to\infty} \frac{4\pi}{T} \mid x_N(\omega) \mid^2 = \frac{4kTD}{m^2} \frac{1}{(\omega^2 - \omega_0{}^2)^2 + (\omega\omega_0/Q)^2} \tag{1.17}$$

The mean square displacement (due to noise) of the mass in an infinitesimal frequency interval $\Delta\nu$ is then

$$\langle x_N{}^2(t) \rangle_{\Delta\nu} = \frac{4kTD\, \Delta\nu}{m^2} \frac{1}{(\omega^2 - \omega_0{}^2)^2 + (\omega\omega_0/Q)^2} \tag{1.18}$$

which is sharply peaked at $\omega = \omega_0$.

We can try something else in an effort to improve the signal to noise ratio. Suppose we choose the resonant frequency of the mass and spring to be different from the frequency of interest and only record the displacement at frequencies very near the frequency of interest. Then according to Eq. (1.18) the noise will be very small. Unfortunately according to (1.2) the signal will also be very small. The general signal to noise ratio measured over a frequency band $\Delta\nu$ centered at ω can be obtained from Eqs. (1.2), (1.18), and (1.5)

$$(S/N)^2 = \frac{\frac{1}{2} \mid x_0 \mid^2}{\langle x_N{}^2(t) \rangle_{\Delta\nu}} = \frac{\omega_0}{4Q\, \Delta\nu} \cdot [(S/N)^2 \text{ in } (1.5)]. \tag{1.19}$$

The factor $\omega_0/4Q\, \Delta\nu$ can be made much greater than unity and so it looks as if this is a better way to measure the displacement. In fact since we are no longer using resonance there is no need for the mechanical amplifier at all; we could just measure the motion $y_0(t)$ directly! There are two major problems with doing the measurement directly although it may indeed be possible to do it directly with a similar signal to noise ratio as above.

One problem is that we have not really said yet how the actual measurement would be done; we have only calculated the signal to noise ratio *inherent* in a certain type of mechanical amplifier. How, exactly, could we measure $x(t)$ [or $y(t)$]? One (good) way would be to convert the motion $x(t)$ [or $y(t)$] to a proportional voltage by some device. For instance the mass could be a conductor and could constitute one plate of a parallel plate capacitor. The motion of the mass would change the gap of the capacitor and thus change the value of its capacitance. This capacitor would be incorporated in one arm of an electrical bridge network and the motion of the mass would change the balance of the bridge. The unbalance signal could then be amplified with an ordinary electric signal amplifier, rectified, averaged by an RC filter, and recorded on a paper chart recorder (Fig. 1.2). Now electrical circuits also have noise associated with them. (Indeed, it was for these that most of noise theory was originally worked out.) We can see this quite easily by noting that the electrons in a circuit have $\frac{1}{2}\,kT$ in each

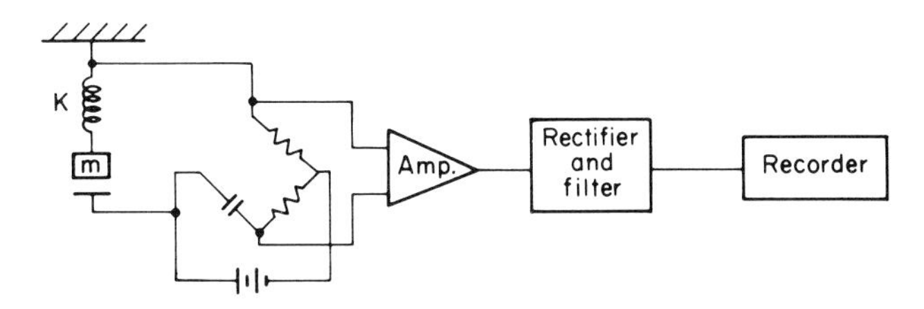

FIG. 1.2. An electromechanical detection system.

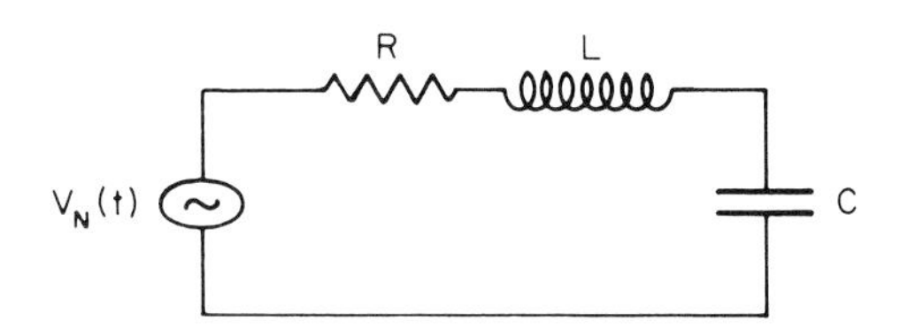

FIG. 1.3. A series LRC circuit.

degree of freedom and therefore there will be random noise currents generated in any circuits. For instance for a series LRC circuit (Fig. 1.3) we have an equation analogous to (1.6)

$$L\ddot{q} + R\dot{q} + \frac{1}{C}\,q = V_N(t) \tag{1.20}$$

where V_N is the equivalent noise voltage generated by the circuit. This circuit has one normal mode (resonance) and therefore there is $\frac{1}{2}\,kT$ associated with the "kinetic energy", $\frac{1}{2}L\langle\dot{q}^2\rangle$, and $\frac{1}{2}\,kT$ with the potential energy, $\frac{1}{2}(1/C)\langle q^2\rangle$. The analysis is identical to that of the mechanical oscillator and so we can immediately write down the amount of noise generated by the resistance R in a band width $\Delta\nu$,

$$\langle {V_N}^2\rangle_{\Delta\nu} = 4kTR\,\Delta\nu\,. \tag{1.21}$$

This is the famous Johnson noise formula and it says that any resistance has associated with it a noise voltage generator and has the equivalent circuit of Fig. 1.4. A more general statement is that, for any complex impedance Z, there is always a noise voltage source in series with Z and the power spectrum of the source is $4kT[\mathrm{Re}(Z)]$.

A consequence of all this is that *additional* noise is generated in the electrical circuits. The reason for using a mechanical amplifier was to get a large enough motion (and associated noise) so that the electrical circuitry would contribute negligible additional noise. If the mechanical amplifier were not used then the main noise contribution would come from the electrical circuitry and the analysis would have to be redone using (1.21) and similar expressions for the noise generated in transistors or vacuum tubes. It is conceivable that you could get a better signal to noise ratio by doing this but it seems to be a rule that a passive device (a device with no external power source, such as the mass and spring) has less noise than an active one (such as an amplifier).

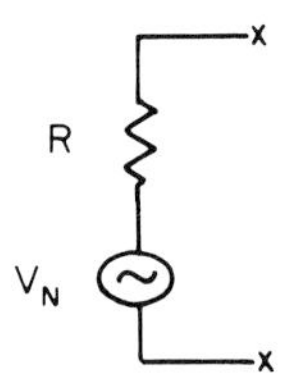

FIG. 1.4. Equivalent circuit for a resistor.

The final point to investigate is that (1.19) indicates that if $Q\ \Delta\nu$ is made small enough you can get a better signal to noise ratio by working off resonance. The reason this will not work can best be described by a more familiar analogy. Suppose you want to measure the average number of particles per second given off by a radioactive material. What you do is count the particles for a time, T, and then divide the total number of counts, N, by T to get the average rate. You also know that the error in N is $\sqrt{N}$ so that the signal to noise ratio for this experiment is just

$$S/N = \frac{N}{\sqrt{N}} = \sqrt{N} = (r\mathrm{T})^{1/2} \tag{1.22}$$

where r is the average counting rate. That is, the signal to noise ratio is proportional to the square root of the time over which you take data. Now for our mechanical circuit we can make the following argument. We found the characteristic buildup and decay time of the mechanical oscillator to be $2Q/\omega_0$. That is, the amplitude of the noise motion of the mass varies in a random manner but it takes approximately $2Q/\omega_0$ sec (on the average) for the amplitude to change appreciably (Fig. 1.5). We can think of these variations (of length $\sim 2Q/\omega_0$) as pulses and ask: How many, N, of these "pulses" occur in a time T?

$$N \approx \frac{\mathrm{T}}{2Q/\omega_0} = \frac{\omega_0 \mathrm{T}}{2Q} \equiv \frac{\Delta\omega \mathrm{T}}{2} = \pi\, \Delta\nu_R \mathrm{T} \tag{1.23}$$

where $\Delta\nu_R$ ($\equiv \nu_0/Q$) is the "bandwidth" of the resonance, i.e., the frequency range over which most of the noise is concentrated [as can be seen from (1.9)]. So by averaging for a time T you can get an improvement of the signal to noise ratio by a factor of

$$\frac{N}{\sqrt{N}} \approx (\pi\, \Delta\nu_R \mathrm{T})^{1/2} = \left(\frac{\omega_0 \mathrm{T}}{2Q}\right)^{1/2} \tag{1.24}$$

in the resonant case. A more precise analysis yields the same result except that there is a numerical factor of order unity that multiplies (1.24). The exact factor depends on the shape of the resonance; for our case it is $\sqrt{2}$

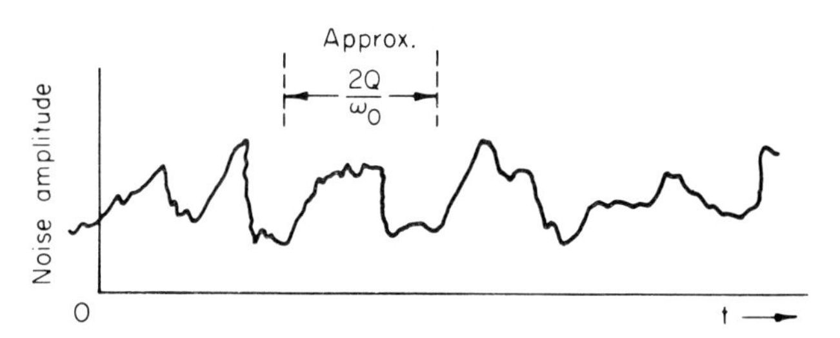

FIG. 1.5. Narrow-band noise.

but we will not worry about this correction. In the off-resonant case the appropriate bandwidth is the $\Delta\nu$ that appears in (1.19). The nonresonant case can be improved by a factor of $\sim (\pi\, \Delta\nu \mathrm{T})^{1/2}$. If we insert these factors into (1.5) and (1.19) we obtain

$$(S/N)^2 = \frac{m\omega_0{}^2 Q^2}{2kT} \frac{\omega_0 \mathrm{T}}{2Q} y_0{}^2 = \frac{m\omega_0{}^3 Q \mathrm{T}}{4kT} y_0{}^2 \qquad \text{on resonance;} \tag{1.25}$$

$$(S/N)^2 = \frac{\pi}{2} \cdot [(S/N)^2 \text{ in } (1.25)] \qquad \text{off resonance.} \tag{1.26}$$

We see then that averaging *always* improves the signal to noise ratio and it also removes the advantage of working off resonance. *It is therefore better to work on resonance* and gain the advantage of the mechanical amplification because the electrical circuitry usually can be designed to contribute a negligible amount of additional noise and therefore (1.25) gives the actual ratio. In the off-resonant case the main noise comes from the amplifiers and so the actual signal to noise ratio would be considerably smaller than (1.26).

Another way to look at it is that the absolute amount of noise contributed by the amplifier is the same in either case. However, on resonance both the mechanical noise and signal are amplified by about the same amount and therefore the signal to *total* noise ratio is larger than if you do not use resonance.

If we use the same numbers in (1.25) as we did in (1.5) and in addition average for about three hours, we could then measure a displacement of about 10^{-11} cm.

We can sum up this problem in three words:

DISSIPATION IMPLIES NOISE!

Since any system has dissipation (with the possible exception of superconductors and superfluids), any system has noise. In many cases all that is necessary is to find the analogous quantities to those in (1.16) or (1.21) to be able to treat the case completely.[1]

[1] A much more thorough treatment of noise in electrical circuits can be found in "Noise" by A. van der Ziel, Prentice-Hall, Englewood Cliffs, New Jersey, 1956.

2. LOW FREQUENCY LOUDSPEAKER DESIGN*

2.1. Problem

Design a loudspeaker for the reproduction of low frequency audible sound, including enclosure to have a flat frequency response below 800 Hz. For the sake of definiteness take:

(a) a 30-cm (12 in.) diameter electromagnetic speaker with a combined motional mass (voice coil and piston) of 10 g and a dc voice coil resistance of 8 Ω. Take the stiffness constant of the piston mounting to be 1.00×10^6 dyn/cm and the frictional resistance to be 1500 g/sec. These values are characteristic of the more compliant high quality units commercially available;

(b) a nonvented simple enclosure of the infinite baffle type;

(c) the unit to be powered by a voltage feedback power amplifier of negligible internal impedance.

2.2. Solution

For theoretical background, nomenclature, units, etc., the reader is referred to standard works.[1-3] The appropriate schematic diagram and equivalent circuits are shown in Fig. 2.1, where

R_0 = "blocked" resistance of voice coil (Ω)
L_0 = "blocked" inductance of voice coil (H)
N = number of turns of voice coil winding
B = magnetic flux density in gap (G)
r = radius of voice coil (cm)
$L = m_c + m_p$ mass of voice coil plus mass of piston (g)
R = mechanical frictional resistance (g/sec)

[1] I. Estermann, ed., "Methods of Experimental Physics," Vol. 1, Part 5. Academic Press, New York, 1959.

[2] H. F. Olsen, "Acoustical Engineering." Van Nostrand, New York, 1957.

[3] L. E. Kinsler and A. R. Frey, "Fundamentals of Acoustics." Wiley, New York, 1950.

* Problem 2 is by W. F. Hornyak.

$C = 1/\gamma$, $\gamma = \gamma_1 + \gamma_2$ combined stiffness constant of locating spider and piston surround (dyn/cm)

$\dfrac{1}{\omega C_v} = \dfrac{\rho c^2(\pi a^2)^2}{\omega V}$ volume reactance at piston (g/sec).

$R_a + j\omega L_a$ = radiation impedance of Rayleigh piston (g/sec).

The radiation impedance for a Rayleigh piston of radius a is[4]

$$R_a = \rho c \pi a^2 R_1(x), \qquad \text{where} \qquad R_1(x) = 1 - \frac{2J_1(x)}{x},$$

$$\omega L_a = \rho c \pi a^2 X_1(x), \qquad \text{where} \qquad X_1(x) = \frac{2S_1(x)}{x},$$

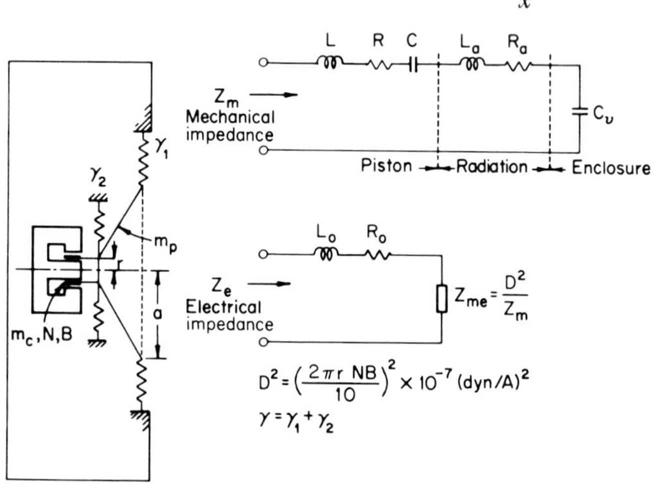

FIG. 2.1. Schematic diagram and equivalent circuits for loudspeaker.

with

ρ = density of air = 1.2×10^{-3} g/cm^3

C = velocity of sound in air = 3.43×10^4 cm/sec

$x = \dfrac{4\pi a}{\lambda}$; λ = wavelength.

For $x < 1$

$$R_1(x) = \frac{x^2}{8} - \frac{x^4}{192} + \cdots$$

[4] For the Struve function $S_1(x)$ and Bessel function $J_1(x)$, refer for example to P. R. E. Jahnke and F. Emde, "Tafeln hoherer Funktionen," 5th ed., pp. 157 and 219. Leipzig, 1952.

$$X_1(x) = \frac{4}{\pi}\left(\frac{x}{3} - \frac{x^3}{45} + \cdots\right).$$

Thus

$$Z_m = R_a + R + j\left[\omega(L_a + M) - \frac{\gamma}{\omega}\right]$$

$$Z_{me} = D^2/Z_m = \frac{D^2}{|Z_m|^2}\left\{(R_a + R_c) - j\left[\omega(L_a + M) - \frac{\gamma}{\omega}\right]\right\}$$

$$Z_e = R_0 + j\omega L_0 + Z_{me}.$$

Clearly if the power amplifier is delivering a terminal voltage E (rms) the average radiated power (watts) is

$$P = \frac{E^2}{|Z_c|^2}\frac{D^2R_a}{|Z_m|^2}$$

at an efficiency of

$$\eta = \frac{D^2R_a}{|Z_m|^2}\Bigg/\left[R_0 + \frac{D^2(R_a + R)}{|Z_m|^2}\right] = \frac{D^2R_a}{D^2(R_a + R) + R_0|Z_m|^2}.$$

At the series resonance of the mechanical circuit $\omega_0{}^2 = \gamma'/(L_a + M)$, with $\gamma' = \gamma_1 + \gamma_2 + [\rho c^2(\pi a^2)^2]/V$, and Z_m (at res.) $= Z_m{}^0 = R_a{}^0 + R$. Further, if R_0 and $\omega_0 L_0$ are small enough (as is generally the case) $Z_e{}^0 \approx Z^0_{me} \approx D^2/(R_a{}^0 + R)$ and hence

$$P^0 \approx \frac{E^2}{D^2}R_a{}^0 \qquad \text{and} \qquad \eta^0 \approx \frac{R_a{}^0}{(R_a{}^0 + R)}.$$

While at $\omega \gg \omega_0$ but still $X = (2\omega a/C) < 1$, the mechanical system is mass dominated and

$$Z_m{}^\omega \approx j\omega(L_a + M), \qquad Z_e{}^\omega \approx R_0, \qquad R_a{}^\omega \approx R_a{}^0\frac{\omega^2}{\omega_0{}^2}.$$

Therefore

$$P^\omega = \frac{E^2D^2R_a{}^0}{\omega_0{}^2(L_a + M)R_0{}^2} \qquad \text{and} \qquad \eta^\omega = \frac{D^2R_a{}^0}{R_0\omega^2(L_a + M)^2}.$$

Thus for $\omega \gg \omega_0$, both P^ω and η^ω are seen to be largely independent of the frequency.

For flat radiated power response with a constant voltage generator set

$$P^0 = P^\omega, \qquad \text{or} \qquad \omega_0(L_a + M)R_0 = D^2,$$

and

$$\eta^\omega = R_a{}^0/\omega_0(L_a + M).$$

Application to a 30-cm diameter speaker (effective $a = 15$ cm) is

$$L_a \text{ (low frequencies)} = \tfrac{8}{3}\rho a^3 = 10.9 \text{ g}$$

$$R_a{}^0 \text{ (at resonance)} = \frac{\pi\rho a^4 \omega_0{}^2}{2c} = 0.00281\ \omega_0{}^2 \text{ g/sec.}$$

With $M = 10$ g if the low frequency limit, generally about f_0, is to be 20 Hz, η^ω is seen to be $\approx 1.8\%$. Thus the present "flat" design is seen to be an inherently low efficiency design; such units nonetheless are currently en vogue. As a reasonable compromise between a desire for a low f_0 and a high η^ω, consider $f_0 = 70$ Hz, then $\eta^\omega = 0.059$. For an average of 50 mW of radiated acoustical power the amplifier is required to supply 0.85 W and for musical reproduction should probably have to have a peak power capacity of 20–30 W.

Then $R_a{}^0$ (at 70 Hz) $= 542$ g/sec and with $R = 1500$ g/sec and $R_0 = 8\ \Omega$,

$$\eta^0 = 0.266$$

$$D^2 = 9200 R_0 = 7.35 \times 10^4 \quad (\text{dyn/A})^2$$

$$\gamma' = 4.04 \times 10^6 \quad (\text{dyn/cm}).$$

The enclosure stiffness is then 3.04×10^6 dyn/cm and hence has a volume

$$3.04 \times 10^6 = \frac{\rho c^2 (\pi a^2)^2}{V} \qquad \text{or} \qquad V = 2.34 \times 10^5 \text{ cm}^3$$

(e.g., 40 cm × 63 cm × 84 cm).

Note that if δ and σ are the density and resistivity of the voice coil wound in N turns of wire of cross sectional area A

$$m_c = 2\pi r N\, \delta A$$

and therefore

$$R_0 = 2\pi r \frac{N\sigma}{A} \qquad R_0 m_c = (2\pi r N)^2\, \delta\sigma$$

which, combined with the definition of D^2, gives

$$B^2 = \frac{\delta\sigma D^2 \times 10^9}{R_0 m_c}.$$

Thus if the material of the voice coil (e.g., copper or aluminium, etc.) is specified as well as R_0 and m_c, the required magnetic field on the mechanical and electrical properties are given and independent of r, N, and A.

For a copper voice coil $\delta = 8.90\ \text{g/cm}^3$ and $\sigma = 1.72 \times 10^{-6}\ \Omega\text{-cm}$, if $m_c = 2$ g and $R_0 = 8\ \Omega$,

$$B = 8400\ \text{G}.$$

Such a coil might be expected to have an inductance of 4×10^{-4} H, thus

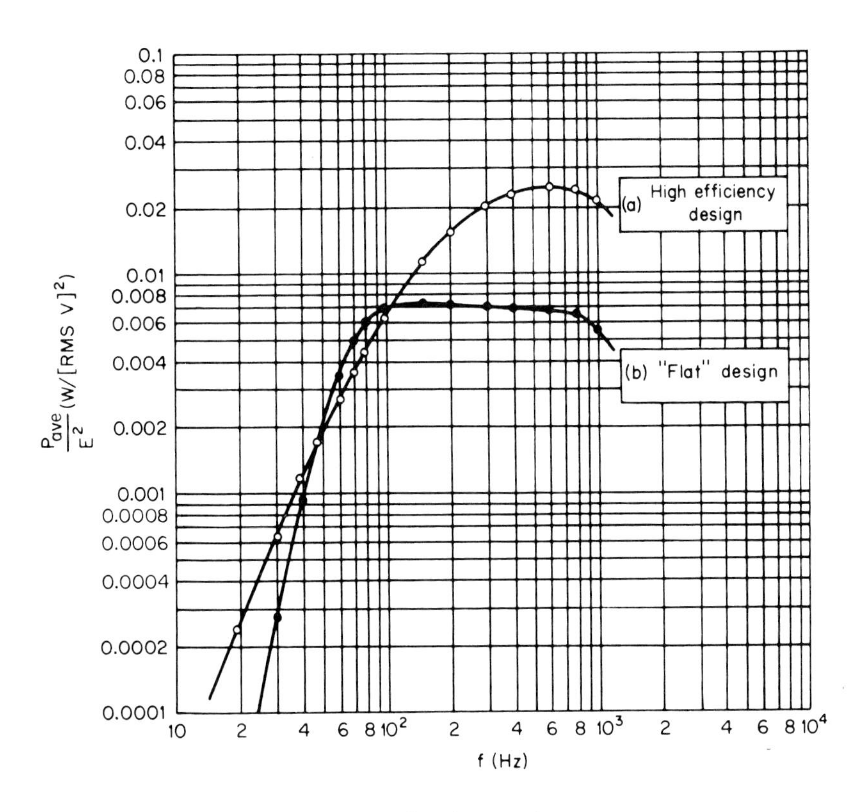

FIG. 2.2. Power radiated versus frequency.

ωL_0 even at 800 Hz is only 2 Ω and can be neglected. The curves given in Figs. 2.2–2.4 show average power radiated, $|Z_e|$, and efficiency as a function of frequency.

The peak voice coil excursion to radiate any given amount of power can be readily calculated since $P = \frac{1}{2}V_0{}^2R_a = \frac{1}{2}\omega^2X_0{}^2R_a$ where U_0 and X_0 are the peak velocity and displacement of the piston. Thus to radiate 50 mW at 70 Hz gives $X_0 = 0.098$ cm.

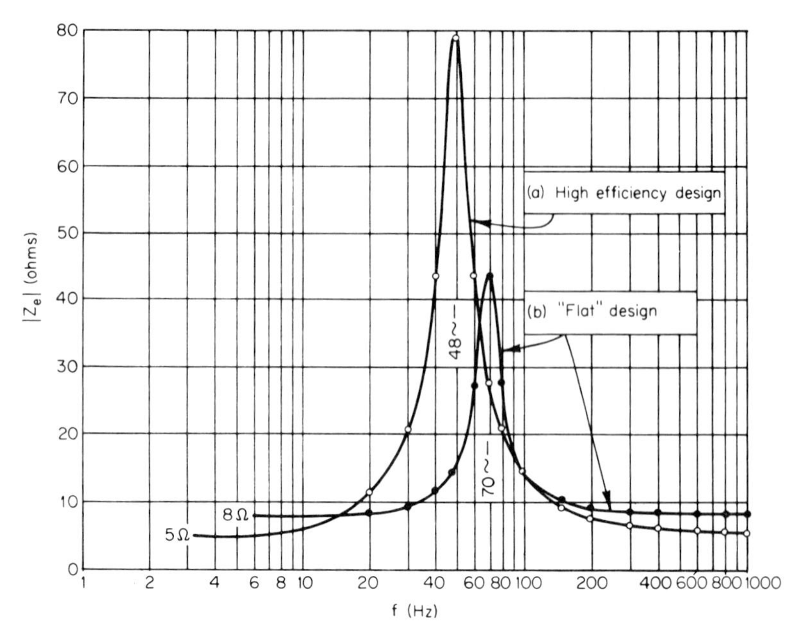

FIG. 2.3. $|Z_e|$ (ohms) versus frequency.

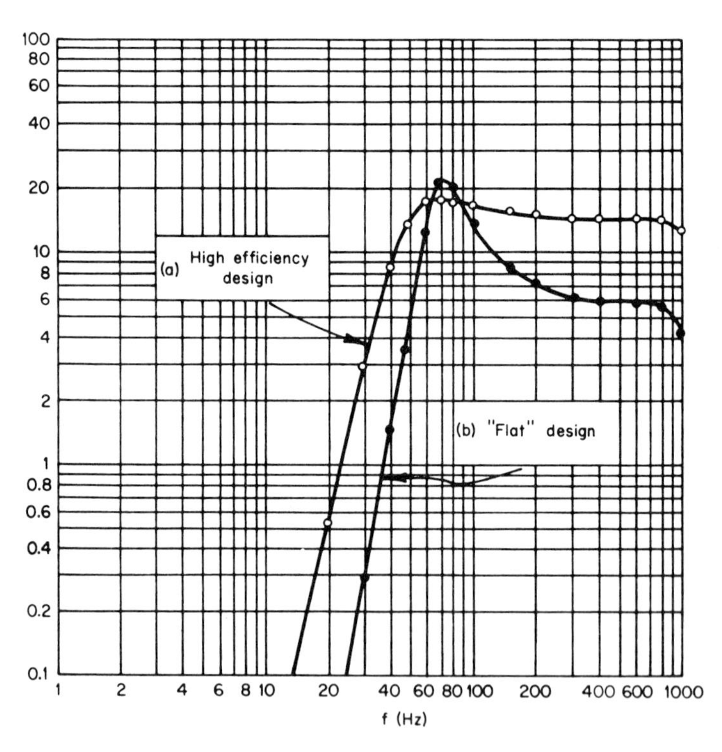

FIG. 2.4. Efficiency versus frequency.

2.3 Exercises

1. Repeat flat design calculations for $f_0 = 20$ Hz.

2. Repeat calculations for a high efficiency design, using the same values for R, M, m_c, and R_0 but allowing D^2 to be as large as possible by taking $B = 15{,}000$ G and an aluminium voice coil. Compare radiated power and efficiency with flat design.

3. Compare the damping (i.e., Q) of the mechanical series resonance for the given design and that of question 2, being sure to include the effect of R_0 reflected into the mechanical circuit.

4. Describe how by noting the coil displacement under an impressed dc voltage and by noting the variation of the resonance with added masses to the piston, one may experimentally determine D^2, M, and γ'. Describe nondestructive tests for determining $\gamma_1 + \gamma_2$, L_a, L_0, R_0, and R_a.

3. TEMPERATURES BELOW 4.2°K*

3.1. Problem

Temperatures in the range of 1–4.2°K are readily produced by pumping on liquid helium. We wish to consider the design of a pumping system capable of maintaining a temperature of 1.5°K or above when the rate of evaporation of liquid helium from our dewar system due to thermal leakage into the dewar is 50 cm^3/h.

3.2. Solution

The pumping system consists of a mechanical pump and appropriate connecting tubes (Fig. 3.1). The solution to our design problem amounts to an intelligent choice of capacity rating of the mechanical pump as well as suitable length and diameter of the interconnecting tubing.

The vapor pressure of liquid He at 1.5°K is found from a table[1] as 3.6 mm of Hg. We wish to maintain P_1, the pressure in the dewar, at this value. The loss of liquid at a rate of 50 cm^3/h implies the need to remove a corresponding amount of gas at the top of the dewar vessel. The top of the dewar will be at room temperature $T \approx 300°K$. The density of liquid He is 0.122 g/cm^3, so that the mass of He to be removed is 6.1 g/h. With reference to STP conditions, we find the volume of gas to be removed per second to be

$$S = \frac{6.1}{8} 22.4 \frac{300}{273} \frac{760}{3.6} \frac{1}{3600} \quad \text{liters/sec}$$

$$= 1.1 \quad \text{liters/sec.}$$

A mechanical pump with a capacity of 1.1 liters/sec or, equivalently, 66 liters/min, if placed immediately at the top of the dewar, would prove sufficient to maintain the desired temperature.

Needless to say, mounting a pump at the dewar head will generally prove impossible. We need to use some finite length of tubing to connect

[1] "Experimental Cryophysics," (F. E. Hoare, L. C. Jackson, N. Kurti, eds.). Butterworths, London, 1961.

* Problem 3 is by J. F. Koch.

the pump to the dewar. Very qualitatively, the effect of the pumping line will be to reduce the pressure at the pump end, so that the pump speed, necessary to remove He gas from the dewar at the same rate, will be increased. For the range of pressures of interest in this problem, the mean

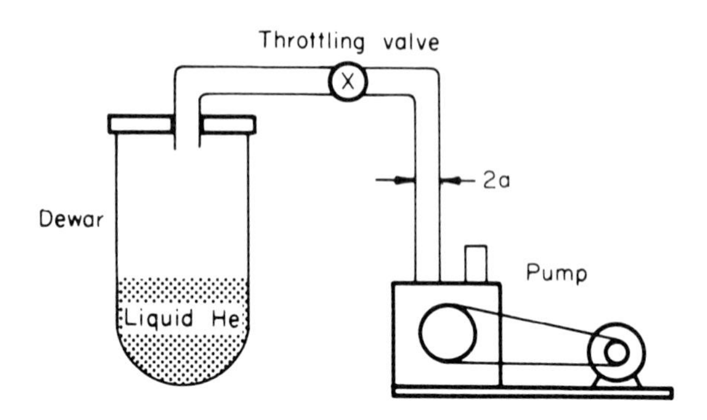

FIG. 3.1. Pumping system.

free path of gas molecules is much less than typical dimensions of the connecting tubing. The decrease in pressure is related to the flow rate at the end of the tube[2] by

$$S_0 = \frac{\pi a^4}{8\eta l}\left(\frac{P_1^{\,2} - P_0^{\,2}}{2P_0}\right).$$

S_0 and P_0 are, respectively, flow rate and pressure at the pumping end of the line; η is the viscosity of the gas; a and l the radius and length of the line. This formula is derived on the assumption of simple laminar flow of a viscous, compressible fluid through a cylindrical tube.

In practice it is desirable to install tubing of sufficient radius so that the resulting decrease in the pressure is small compared to P_1. When this is the case, $P_1 \approx P_0$ and

$$S_0 \approx \frac{\pi a^4}{8\eta l}(P_1 - P_0).$$

In any case, when the pressure at the end of the tube is P_0, the capacity of the pump will have to be increased to

$$S_0 = S\left(\frac{P_1}{P_0}\right)$$

where S is the flow rate at the dewar end.

[2] See, for example, A. Sommerfeld, "Mechanics of Deformable Bodies," Academic Press, New York, 1950.

For our problem let us assume that we are using a 1-in. diameter line, 2 m long. Using $\eta = 189 \times 10^{-6}$ cgs units, $a = 1.27$ cm, $l = 200$ cm, the pressure difference expressed in millimeters of Hg becomes

$$(P_1 - P_0) \approx \frac{1.1 \times 10^3 \times 8 \times 189 \times 10^{-6} \times 200}{3.14 \times (1.27)^4} \left(\frac{1}{1.36 \times 980}\right)$$

$$\approx 0.3 \quad \text{mm of Hg.}$$

The capacity rating of the pump will have to be increased by about 9% to ~ 72 liters/min.

The system of line and mechanical pumps, with the above specifications, will be capable of maintaining a temperature of 1.5°K in our dewar system. An additional throttling valve in the pumping line will serve to decrease further the pumping speed of the line and achieve temperatures in the range 1.5–4.2°K.

The problem as posed was somewhat arbitrary with respect to choice of the pumping line dimensions, because we assumed that we had an unlimited choice of pumps. More realistically, the experimenter will have available a single pump with a certain capacity rating. Also, the length of the line will be determined by physical dimensions and arrangement of his experimental equipment. Under these conditions there would be a unique diameter of tubing that would achieve the desired pressure.

3.3. Exercises

1. If the experimenter has available a 100-liter/min pump, what will be the minimum pressure that can be maintained in the dewar with the line chosen in the problem?
2. If a 1.25 cm diameter line had been chosen, what would be the required capacity rating of the pump?

4. REFLECTIVITY MEASUREMENT*

4.1. Problem

Design a setup to detect changes in the reflectivity, at the optical range of 4500–5500 Å, of a metal surface to one part in a 100,000 for near normal incidence. The reflectivity change to be measured is produced by applying a magnetic field of about 50 G. State the requirements of the source, type of detector, and give a block diagram of the apparatus.

4.2. Solution

There are two important considerations to be taken in account in order to measure intensity changes of one part in 10^5. These are the limit due to statistical fluctuations of the number of photons being detected, and limitations of the experimental technique.

The reflectivity is to be measured at the wavelength interval 4500–5500 Å. A phototube or photomultiplier tube is the most efficient detector in this range.[1,2] The rms noise in the photocurrent of a photomultiplier or phototube is given by[3]

$$i_n = (2ie \, \Delta f)^{1/2}$$

where i_n is the rms photocurrent noise at the cathode, i is the average photocurrent at the cathode, e is the electronic change, and Δf is the bandwidth of the detecting apparatus. Choosing a representative bandwidth of $\Delta f = 1$ Hz, we find a signal to noise ratio of

[1] R. W. Engstrom, *in* "Methods of Experimental Physics" (E. Bleuler and R. O. Haxby, eds.), Vol. 2, Chapter 11.1. Academic Press, New York, 1964.

[2] I. Ames and R. L. Christensen, *in* "Methods of Experimental Physics" (V. W. Hughes and H. L. Schultz, eds.), Vol. 4A, Section 1.1.3. Academic Press, New York, 1967.

[3] K. M. van Vliet, *in* "Methods of Experimental Physics" (E. Bleuler and R. O. Haxby, eds.), Vol. 2, Part 12. Academic Press, New York, 1964.

* Problem 4 is by E. A. Stern and W. F. Hornyak.

$$\frac{S}{N} = \frac{i}{i_n} = \left(\frac{i}{2e}\right)^{1/2} = \left(\frac{n}{2}\right)^{1/2}$$

where S/N is the signal to noise ratio and n is the number of photoelectrons emitted from the photocathode per second.

In the wavelength range considered photocathode surfaces have a quantum efficiency of about 10% so that about 10 photons are required to produce one photoelectron. From the requirement of the problem that $S/N \geqq 10^5$ and that $\hbar\omega$, the energy of one photon of 5000-Å wavelength, is about 2.5 eV, one finds that the required light intensity I at the photomultiplier is $I \geqq 10n\ \hbar\omega \approx 0.8$ erg/sec $= 0.8 \times 10^{-7}$ W. This is a very feasible intensity since the total emissive power from a tungsten filament at 2450°K is 50 W/cm^2. A tungsten lamp powered by a constant current source such as a storage battery is a good choice for the light source since sufficient intensity can be obtained and it also has the required stability of emitted light.

To discriminate a light intensity change of one part in 10^5 the magnetic field on the reflecting surface can be varied between 0 and 50 G at a fixed frequency in the range of, say, 10–1000 Hz. By measuring the sinusoidal component, only the change in intensity due to changes in reflectivity is detected. To minimize effects of fluctuations of the light source intensity, the light before reflecting can be split and observed by one photomultiplier or phototube while the reflected light is observed by another. The difference in output between the two phototubes can then be taken tending to cancel fluctuations of the light source but not affecting the measurement of the change in reflectivity.

4.3. Exercises

1. What considerations enter in choosing the frequency at which the magnetic field is varied?
2. The fluctuating magnetic field can directly vary the output of the phototubes. What can one do to make this effect negligible?
3. What considerations enter in choosing whether to detect the signal with a photomultiplier tube or phototube?

A possible experimental setup is shown in Fig. 4.1.

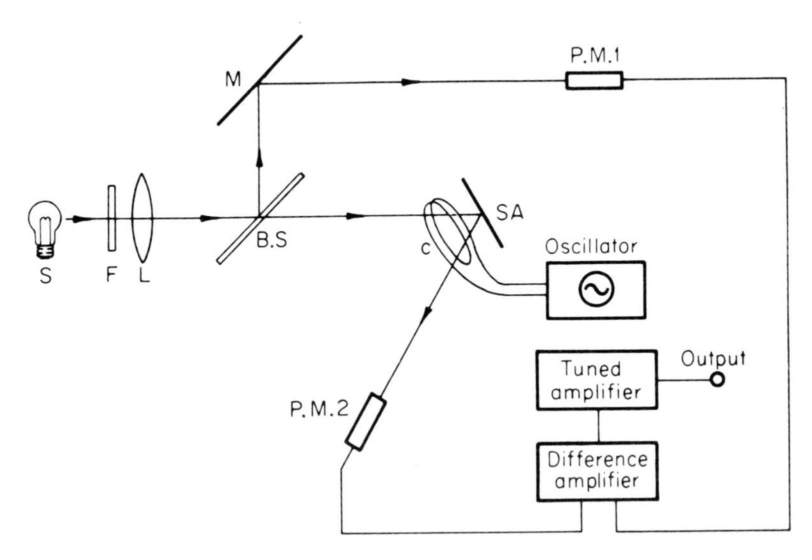

FIG. 4.1. Block diagram of possible experimental setup where: S is the tungsten lamp light source; F is a 4500–5500 Å light filter; L is a condensing lens; B.S. is a beam splitter; M is a mirror; SA is the sample under study; C is a coil producing modulating magnetic field; P.M.1 is a phototube or photomultiplier; and P.M.2 is a phototube or photomultiplier.

5. CALIBRATION OF A SPECTROSCOPIC SYSTEM*

5.1. Problem

In many experimental situations it is necessary to make absolute measurements of light intensity in order that collision cross sections and other such fundamental parameters may be determined absolutely. By what means can the experimenter calibrate his light detection system with relatively simple equipment while maintaining good reproducibility and accuracy?

5.2. Solution

Radiant energy measurement divides itself into two classes: the first is the measurement of the total radiant energy or flux (i.e. for all wavelengths) from the source, and the other is the measurement of radiant energy at a particular small wavelength interval. This paper is directed toward the latter type of measurement. A standard source is used for calibration of the detection system. There are many sources of radiation,[1] some having continuous spectra such as the blackbody, incandescent lamp, carbon arc, and globar. Prime considerations are reproducibility, ease of operation, and desirable spectral distribution. A blackbody is considered the ultimate standard, as its emission can be accurately calculated by use of Planck's radiation law. However, it is rather difficult to set up and use. The arcs and glows are dependent upon many parameters and less reproducible than the hot bodies. The tungsten ribbon lamp is ideal for meeting the requirements as its radiation departs in a predictable way from blackbody radiation when a properly purified[2] ribbon is used in an evacuated chamber. The following material is dedicated to the use of a tungsten ribbon standard lamp in the measurement of absolute light intensities.

A solid body will emit radiant energy much as does a blackbody. Its radiant flux is that from a blackbody corrected by a factor known as the

[1] W. E. Forsythe (ed.), "Measurement of Radiant Energy," 1st edition, Chapter II. McGraw-Hill, New York, 1937.

[2] R. D. Larrabee, *J. Opt. Soc. Am.* **49,** 619 (1959).

* Problem 5 is by Robert M. St. John.

emissivity which is dependent upon the true temperature of the body and the wavelength of the emission. The emissivity is always less than one; that is, a hot body always emits less radiation than a blackbody operating at the same temperature. Good experimental values of the emissivity have been obtained by Larrabee[2] and DeVos.[3] The data of Larrabee are preferred because of his very careful purification process.

The true temperature of the ribbon is a function of the electric current passing through it and is obtained by comparison with a blackbody. Lamps commercially manufactured can be obtained with calibration measurements traceable to the National Bureau of Standards. The temperature falls off slightly at the edges from that of the central region. Temperatures are reported for the center part of the filament.

5.2.1. Blackbody Theory

Planck's blackbody theory yields an equilibrium energy density $U^B(\lambda, T)\,d\lambda$ in the interior of a cavity at a temperature T for the wavelength interval between λ and $\lambda + d\lambda$ which is

$$U^B(\lambda, T)\,d\lambda = \frac{8\pi hc}{\lambda^5[e^{hc/\lambda kT} - 1]}\,d\lambda. \tag{5.1}$$

In the expression h is Planck's constant, c is the speed of light in a vacuum, and k is Boltzmann's constant.

The hemispherical power radiancy $I_H{}^B(\lambda, T)$ is related to the energy density by expression

$$I_H{}^B(\lambda, T) = \frac{c}{4}\,U^B(\lambda, T) \tag{5.2a}$$

$$= \frac{2\pi hc^2}{\lambda^5}\,\frac{1}{(e^{hc/\lambda kT} - 1)} \tag{5.2b}$$

$$= \frac{c_1}{\lambda^5}\,\frac{1}{(\exp(c_2/\lambda T) - 1)}. \tag{5.2c}$$

The dimensions of $I_H{}^B(\lambda, T)$ are power per area per unit wavelength radiated into the hemisphere outward from a plane element of the radiating surface. The values of the radiation constants adopted by the National Bureau of Standards in 1963[4] are:

$$c_1 = 3.7405(\pm 3) \times 10^{-16} \quad \text{W–m}^2$$

and

$$c_2 = 1.43879(\pm 19) \times 10^{-2} \quad \text{m–°K}.$$

[3] J. C. DeVos, *Physica* **20,** 715 (1954).

[4] Nat. Bur. Stds. Tech. News Bull., Oct. 1963.

For wavelengths less than that at the peak of the density curve ($\lambda_{max}\ T = 2.89 \times 10^7 \text{Å–°K}$) Planck's law can be simplified to Wien's law which is given by

$$I_H{}^B(\lambda, T) = \frac{c_1}{\lambda^5} \exp\left(\frac{-c_2}{\lambda T}\right). \tag{5.3}$$

In as much as light emitted into a small solid angle usually is measured, a directional radiancy expression is desirable. The directional radiancy for light emitted along the normal is $I_N{}^B(\lambda, T)$ while that for light emitted in a direction θ away from the normal is $I_\theta{}^B(\lambda, T)$. The dimensions are power per area unit wavelength per unit solid angle. Lambert's law for radiation

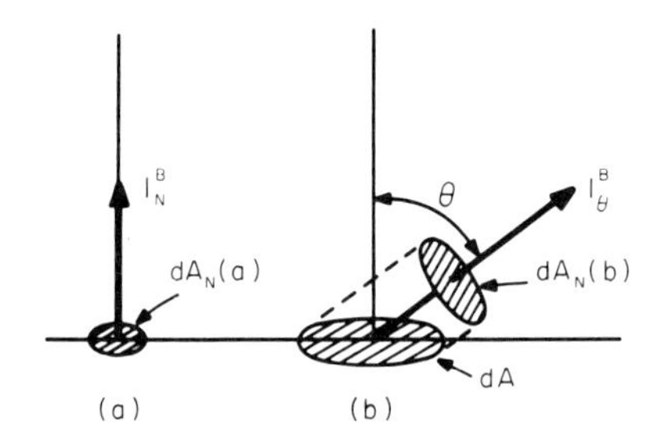

FIG. 5.1. Illustration of Lambert's law. In (a) and (b) the radiated flux is equal provided $dA_N(a) = dA_N(b)$.

states that the brightness of a blackbody is independent of the direction from which it is observed. This means that the power radiated from elements of area with different angles θ are equal provided that the elemental areas all have equal values when projected upon the direction of radiation, i.e. when the dA_N values are equal. See Fig. 5.1. If we let

$$I_\theta{}^B(\lambda, T) = I_N{}^B(\lambda, T) \cos\theta \tag{5.4}$$

it is evident that Lambert's law is fulfilled as

$$I_\theta{}^B(\lambda, T)\, dA = I_N(\lambda, T) \cos\theta\, dA = I_N(\lambda, T)\, dA_N. \tag{5.5}$$

The relationship between $I_N{}^B(\lambda, T)$ and $I_H{}^B(\lambda, T)$ is obtained by integrating $I_\theta{}^B(\lambda, T)$ over the hemisphere. Thus:

$$\int_0^{2\pi} I_\theta{}^B(\lambda, T)\, d\Omega = \int_0^{\pi/2} I_N{}^B(\lambda, T) \cos\theta\, (2\pi \sin\theta\, d\theta) = \pi I_N{}^B(\lambda, T). \tag{5.6}$$

But

$$\int_0^{2\pi} I_\theta{}^B(\lambda, T)\, d\Omega = I_H{}^B(\lambda, T). \tag{5.7}$$

So

$$I_N{}^B(\lambda, T) = \frac{1}{\pi} I_H{}^B(\lambda, T). \tag{5.8}$$

5.2.2. Nonblackbody Modifications

The radiancy expressions for a nonblackbody are obtained by multiplying the blackbody radiance by the emissivity $e(\lambda, T)$. Thus,

$$I_N(\lambda, T) = e(\lambda, T)\, I_N{}^B(\lambda, T). \tag{5.9}$$

In atomic physics experiments the rate of events being measured is determined by the rate of emission of photons from the reaction zone because of the one-to-one relationship between the atomic events and the photons emitted. Each photon of frequency ν carries an energy of $h\nu = hc/\lambda$. The division of the power radiancy expressions by $h(c/\lambda)$ yields photon radiancy expressions. Thus,

$$R_N(\lambda, T) = I_N(\lambda, T)\, \frac{\lambda}{hc}. \tag{5.10}$$

This gives the rate of emission of photons per area per unit wavelength interval into a unit solid angle normal to the surface.

5.2.3. Transmissivity of the Optical System

Light emitted from an experimental volume such as the interior of a collision chamber is measured by its comparison with a known light flux from the standard source. Each beam traverses an optical system in going to the detector, part of which is common to both beams. The two optical systems will be divided into three parts, each with a transmissivity which is wavelength dependent. That part of the systems which is mutually shared by each system will have a transmissivity of $\Upsilon_M(\lambda)$. This would include a monochromator or filter and perhaps several other elements such as lenses, mirrors, windows, etc. The section of the optical system exclusively in the beam from the standard lamp has a transmissivity of Υ_{SL} and would include the window in the standard lamp envelop and perhaps other elements. The transmissivity of the optical system exclusively in the beam from the collision chamber is Υ_{cc}. The transmissivity of a particular element is the fraction of the incident radiation of a given wavelength which actually emerges from the element. It is affected by reflection losses as well as absorption losses. Each interface produces a transmissivity equal to unity less the reflectivity. The transmissivity of a group of elements is equal to the product of the transmissivities of the several interfaces and absorptive thicknesses. Each component of the transmissivity is wavelength dependent, but none vary rapidly with wavelength under normal circumstances except that of the monochromator or filter. Since the mutually shared part has a transmissivity which varies with wavelength in a manner that can be determined relatively, but not absolutely (in the case of a monochromator), one may express $\Upsilon_M(\lambda)$ as a product of two factors. These are: (a) Υ_A, which is the absolute peak value of the transmissivity; and (b) $\Upsilon_R(\lambda)$, which is the shape function with a peak value of unity. Thus

$$\Upsilon_M(\lambda) = \Upsilon_A \Upsilon_R(\lambda). \tag{5.11}$$

The Υ_A factor will later cancel when the light from the two sources is compared.

5.2.4. Photon Flux Reaching the Detector

The photons from the standard lamp which reach the detector are emitted from a small area A_{SL} of the standard lamp ribbon. The area A_{SL} is determined by field stops[5] located outside the ribbon chamber and external to the monochromator. The monochromator entrance slit may form all or part of the field stop, however. The solid angle Ω_{SL} of the radiation passing through the entire optical system is limited by the aperture stop.[5] This, too, should be separate from other natural stops, have easy access, and be adjustable for controlling the light flux. Care must be exercised in order that the stop be the aperture stop for entire area A_{SL}. This may be checked by removing the cover from the monochromator and seeing that the light passing through the entrance slit all falls within the bounds of the reflecting mirror and on the active surface of the grating or prism.

The rate at which photons from the standard lamp are incident upon the detector in the wavelength range from λ to $\lambda + d\lambda$ is

$$R_N(\lambda, T) A_{SL} \Omega_{SL} \Upsilon_{SL} \Upsilon_A \Upsilon_R(\lambda). \tag{5.12}$$

The total photon flux incident upon the detector from the standard lamp is F_{SL} where

$$F_{SL} = A_{SL} \Omega_{SL} \Upsilon_{SL} \Upsilon_A \int_{\text{all } \lambda} R_N(\lambda, T) \Upsilon_R(\lambda)\, d\lambda. \tag{5.13}$$

A knowledge of the function $\Upsilon_R(\lambda)$ is necessary before integration is undertaken. This function is governed by the widths of the entrance and exit slits. (See Ref. 6, Ch. 6.) Equal slits yield a triangular pattern while

[5] F. A. Jenkins, and H. E. White, "Fundamentals of Optics," 3rd edition, Chapter 7. McGraw-Hill, New York, 1957.

unequal slits yield trapezoidal patterns as shown in Fig. 5.2. With equal slit widths

$$\Delta\lambda = \frac{w}{F(d\alpha/d\lambda)} \tag{5.14}$$

where w is the slit width, F is the focal length of the instrument, and $d\alpha/d\lambda$ is the angular dispersion. (See Ref. 6, Ch. 2.)

The trapezoidal pattern has a half-intensity bandwidth of $\Delta\lambda_1$, a full bandwidth of $\Delta\lambda_1 + \Delta\lambda_2$, and a bandwidth of the flat top of $\Delta\lambda_1 - \Delta\lambda_2$.

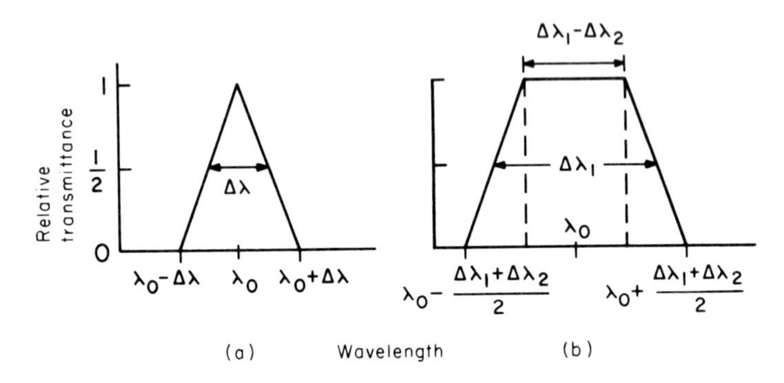

FIG. 5.2. Relative transmittance curves for radiation transmitted by a monochromator with exit slit and entrance slit image (a) of equal width and (b) of different width.

Letting the wider slit be w_1, the narrower slit be w_2, F the focal length of the instrument (same for entrance and exit), it follows that[6]

$$\Delta\lambda_1 = \frac{w_1}{F(d\alpha/d\lambda)}; \qquad \Delta\lambda_2 = \frac{w_2}{F(d\alpha/d\lambda)}. \tag{5.15}$$

It should be noted that for the triangular pattern full bandpass is twice $\Delta\lambda_1$ while for a wide trapezoid the full bandpass is nearly equal to $\Delta\lambda_1$.

For atomic lines, the triangular pattern is often used. The following assumptions are made in order that the integration of Eq. (5.13) can be performed:

A.
$$\lambda = \lambda_0 + x = \lambda_0(1 + x/\lambda_0) \tag{5.16a}$$
$$d\lambda = dx. \tag{5.16b}$$

B.
$$\Upsilon_R(\lambda) = \Upsilon_R(x) = 1 + \frac{x}{\Delta\lambda} \qquad -\Delta\lambda \leqq x \leqq 0 \tag{5.17a}$$
$$= 1 - \frac{x}{\Delta\lambda} \qquad 0 \leqq x \leqq \Delta\lambda. \tag{5.17b}$$

[6] G. R. Harrison, R. C. Lord, and J. R. Loofbourow, "Practical Spectroscopy," Chapters 2, 6. Prentice-Hall, Englewood Cliffs, New Jersey, 1948.

C. The energy density $U^B(\lambda, T)$ varies but slowly over the wavelength range of $\lambda - \Delta\lambda$ to $\lambda + \Delta\lambda$ as shown in Fig. 5.3. It may be expressed as

$$U^B(\lambda, T) = U^B(\lambda_0, T) + \frac{dU^B}{d\lambda} x \tag{5.18a}$$

$$= U^B(\lambda_0, T) + Sx$$

$$= U^B(\lambda_0, T)\left[1 + \frac{xS}{U^B(\lambda_0, T)}\right]. \tag{5.18b}$$

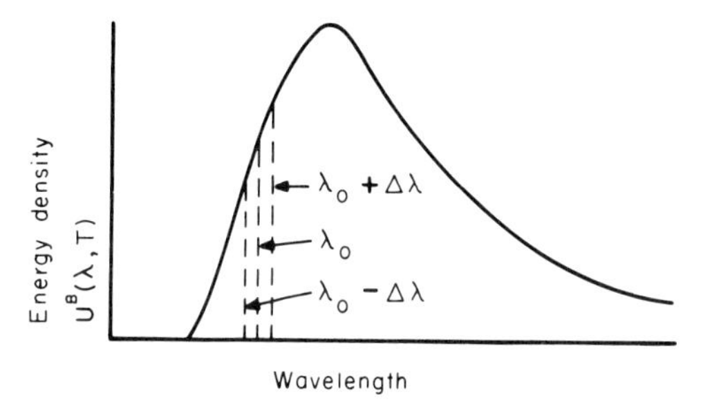

FIG. 5.3. The energy density $U^B(\lambda, T)$ curve showing its variation over the wavelength range from $\lambda_0 - \Delta\lambda$ to $\lambda_0 + \Delta\lambda$.

Here S is the slope of the radiation curve and has the form

$$S = \frac{c_1 \exp(-c_2/\lambda T)}{\lambda^6}\left(\frac{c_2}{\lambda T} - 5\right) \tag{5.19}$$

when the wavelength is on the low side of the peak of the curve (i.e., when Wien's law holds).

D. The emissivity is constant over the wavelength range from $\lambda_0 - \Delta\lambda$ to $\lambda_0 + \Delta\lambda$. Thus

$$e(\lambda, T) = e(\lambda_0, T). \tag{5.20}$$

Letting the integral of Eq. (5.13) be denoted as $R_{TG}(\lambda_0, \Delta\lambda, T)$ for the triangle case it is seen that when T is held constant

$$R_{TG}(\lambda_0, \Delta\lambda, T) = \int_{\lambda_0 - \Delta\lambda}^{\lambda_0 + \Delta\lambda} R_N(\lambda, T)\Upsilon_R \, d\lambda \tag{5.21a}$$

$$= \int_{\lambda_0 - \Delta\lambda}^{\lambda_0 + \Delta\lambda} \frac{e(\lambda, T)}{4\pi h} U^B(\lambda, T)\lambda T_R(\lambda) \, d\lambda \tag{5.21b}$$

$$= \frac{e(\lambda_0, T)}{4\pi h} U^B(\lambda_0, T)$$

$$\times \left| \begin{array}{l} \int_{-\Delta\lambda}^{0} \left[1 + \frac{Sx}{U^B(\lambda_0, T)}\right]\left[1 + \frac{x}{\lambda_0}\right]\left[1 + \frac{x}{\Delta\lambda}\right] dx \\ + \int_{0}^{\Delta\lambda} \left[1 + \frac{Sx}{U^B(\lambda_0, T)}\right]\left[1 + \frac{x}{\lambda_0}\right]\left[1 - \frac{x}{\Delta\lambda}\right] dx \end{array} \right| \tag{5.21c}$$

$$= \frac{e(\lambda_0, T)}{4\pi h} U^B(\lambda_0, T)\lambda_0 \, \Delta\lambda \left[1 + \frac{S(\Delta\lambda)^2}{6U^B(\lambda_0, T)\lambda_0}\right]. \tag{5.21d}$$

The expression $R_{RG}(\lambda_0, \Delta\lambda, T)$ may be evaluated by the use of a modern computer for a variety of values of λ_0, $\Delta\lambda$, and T which are most desirable for the experimental data to be analyzed. The $(\Delta\lambda)^2$ term in the brackets is negligible for usual experimental situations. For example, when $\Delta\lambda = 100$ Å, $\lambda_0 = 3000$ Å, and $T = 1500$°K the correction due this term is only 0.5%. In many cases $\Delta\lambda$ will be much less than 100 Å.

The signal of the detector exposed to the light flux F_{SL} from the standard lamp is proportional to it. Thus

$$F_{SL} = B(\lambda)I_{SL} \tag{5.22}$$

where

$$B(\lambda) = \frac{F_{SL}}{I_{SL}} = \frac{R_{TG}(\lambda_0, \Delta\lambda, T)A_{SL}\Omega_{SL}\Upsilon_{SL}\Upsilon_A}{I_{SL}}. \tag{5.23}$$

The photon flux $F_{cc}(D)$ due to an atomic transition in the collision chamber which reaches the detector also is related to the detector signal by the constant $B(\lambda)$. Thus

$$F_{cc}(D) = B(\lambda)I_{cc}. \tag{5.24}$$

The photon flux $F_{cc}(\Omega_{cc})$ emitted into the optical system is related to $F_{cc}(D)$ by

$$F_{cc}(D) = \Upsilon_{cc}\Upsilon_A\Upsilon_R F_{cc}(\Omega_{cc}) = \Upsilon_{cc}\Upsilon_A F_{cc}(\Omega_{cc}) \tag{5.25}$$

as $\Upsilon_R = 1$ for atomic line transitions. The symbol Ω_{cc} signifies the solid angle of the radiation from the collision chamber which traverses the light detection system. This solid angle is defined by an aperture stop. Again, care must be exercised in order that the stop be the aperture stop for the entire section of the beam viewed in the collision chamber. Furthermore, the total photon flux from the collision chamber $F_{cc}(4\Pi)$ emitted into 4Π sr is related to $F_{cc}(\Omega_{cc})$ by

$$F_{cc}(4\Pi) = \frac{4\Pi}{\Omega_{cc}} F_{cc}(\Omega_{cc}). \tag{5.26}$$

It follows from Eqs. (5.24), (5.25), and (5.26) that

$$F_{cc}(4\Pi) = \frac{4\Pi}{\Omega_{cc}} \frac{1}{\Upsilon_{cc}\Upsilon_A} B(\lambda) I_{cc}(D). \tag{5.27}$$

From Eqs. (5.23) and (5.27)

$$F_{cc}(4\Pi) = 4\Pi A_{SL} R_{TG}(\lambda_0, \Delta\lambda, T) \frac{\Omega_{SL}}{\Omega_{cc}} \frac{\Upsilon_{SL}}{\Upsilon_{cc}} \frac{I_{cc}}{I_{SL}}. \tag{5.28}$$

5.2.5. An Experimental Application

Experiments involving atomic and electronic collisions at the University of Oklahoma require the measurement of radiation emitted from a collision chamber. Figure 5.4 shows the experimental arrangement used. The light

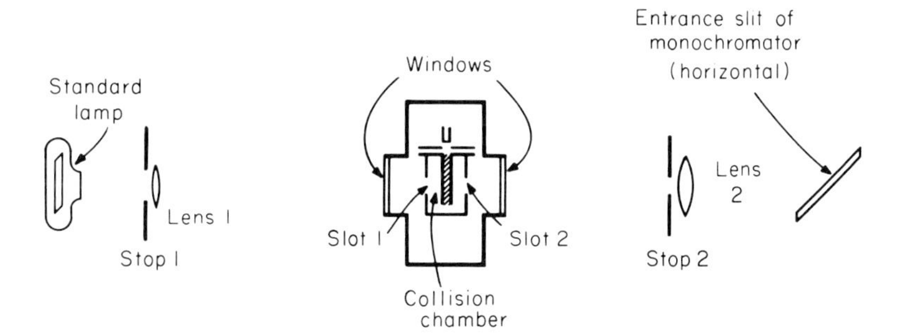

FIG. 5.4. An optical system showing the components for using a tungsten ribbon as a standard for the absolute measurement of light emitted in electronic and atomic collision experiments.

from the standard lamp passes through circular stops 1 and 2, lens 1 and 2, slots 1 and 2 in the collision chamber, and through the monochromator entrance slit enroute to the detector at the exit slit of the monochromator. An image of the standard lamp filament is formed on the front surface of the collision chamber by lens 1 where slot 1 acts as the field stop, always limiting the horizontal spread of the field of view and sometimes limiting the vertical part of this field also. A very narrow entrance slit on the monochromator may act as the vertical field stop, however. The circular stop 1 acts as the aperture stop for rays from the standard lamp. Its diameter and distance from the filament is used in determining the solid angle Ω_{SL}. A magnified image is formed by lens 1 in order that only the center of the filament be viewed by the detector.

The pencil of gas excited by the electron beam in the collision chamber is the second light source. The vertical length of the beam viewed by the detector is limited by slot 2 of the chamber while the entire width of the beam is viewed. The circular stop 2 acts as the aperture stop for the rays from the collision chamber, and is used in the calculation of Ω_{cc}. This stop will not limit the radiation from the standard lamp unless it is made very small. The standard lamp and collision chamber are operated at different times to give the separate detector outputs.

The detector near the exit slit of the monochromator should be large enough to receive all the light passing through the slit as the exit slit is not necessarily uniformly illuminated. A response of the detector uniform over its sensitive area is a second requirement for the calibration to be accurate.

5.2.6. Trapezoidal Monochromator Bandpass

In instances when the experimental light intensity being measured is not monochromatic, but falls within a restricted range as do many molecular bands, a trapezoidal bandpass of the monochromator may be used. The entrance slit should be narrow and the exit slit wide. This will permit all of the light (whose wavelength is within the flat top region of the transmitted wavelength band) which enters the entrance slit to pass through the exit slit and be detected. Thus, a nonuniform illumination of the entrance slit will have had no seriously deleterious effects. A broad entrance slit and narrow slit would produce the same trapezoidal transmittance pattern, but nonuniform illumination of the entrance slit would have serious consequences on the accuracy of the calibration.

If one assumes the trapezoid has very steep sides as when the exit slit is several times wider than the entrance slit, $\Upsilon_R(\lambda)$ can be presented by

$$\begin{aligned} \Upsilon_R(\lambda) &= 1 \qquad -\Delta\lambda_1 \leqq x \leqq \Delta\lambda_1 \\ &= 0 \qquad x < -\Delta\lambda_1; \quad x > \Delta\lambda_1. \end{aligned} \tag{5.29}$$

The integral of Eq. (5.13) for the trapezoidal bandpass is

$$R_{TZ}(\lambda_0, \Delta\lambda_1, \Delta\lambda_2, T) = \frac{e(\lambda_0, T)}{4\Pi h} U^B(\lambda_0, T)\lambda_0 \, \Delta\lambda_1 \left[1 + \frac{1}{18} \frac{S(\Delta\lambda_1)^2}{U^B(\lambda_0, T)\lambda_0}\right]. \tag{5.30}$$

Again, the $(\Delta\lambda_1)^2$ term in the brackets is usually negligible when $\Delta\lambda \leqq$

100 Å. It is to be noted that the simplified form of $R_{TZ}(\lambda_0, \Delta\lambda_1, \Delta\lambda_2, T)$ is equal to that of $R_{TG}(\lambda_0, \Delta\lambda, T)$.

A trapezoidal pattern in which $\Delta\lambda_1$ and $\Delta\lambda_2$ are of comparable values also yields a radiancy of

$$R_{TZ}(\lambda_0, \Delta\lambda_1, \Delta\lambda_2, T) = \frac{e(\lambda_0, T)}{4\Pi h} U^B(\lambda_0, T)\lambda_0 \, \Delta\lambda_1 \qquad (5.31)$$

where the $\Delta\lambda^n$ terms such as appeared in the bracketed multipliers of Eqs. (5.19d) and (5.28) are neglected.

All the above expressions for the integrated radiancy are proportional to the area of the curve bound by $\Upsilon_R(\lambda)$. The expressions all are proportional to $\Delta\lambda_1$, the half-intensity bandwidth. For the triangle $\Delta\lambda_1$ is one-half of the full bandwidth and for the rectangle it is the full bandwidth.

The advantage of the use of an exit slit far exceeding the entrance slit is that all the light within a band emitted from the collision chamber is transmitted uniformly to the detector.

6. VISIBILITY OF YOUNG'S INTERFERENCE FRINGES*

6.1. Problem

In all of the problem it will be assumed that the source emits monochromatic radiation of wavelength $\lambda = 0.55\ \mu\text{m}$.

6.1.1.

6.1.1.1. A point source S_0 illuminates two vertical identical slits F_1 and F_2 in an opaque screen. The slits are narrow, parallel, and separated by the distance $s = 2$ mm. The interference effects are observed in a plane π parallel to the plane of the slits and at a distance $D = 1$ m from this plane. Any point M in the plane π will be designated by its coordinates X and Y (Y parallel to the slits). Compute and show graphically the manner in which the illumination varies in the plane π.

6.1.1.2. How is this illumination changed if S_0 is replaced by a narrow slit F_0 parallel to F_1 and F_2? Determine the fringe pattern.

6.1.1.3. For observing the fringes a Fresnel lens comparable to a thin lens of focal length 2 cm is used. What advantage results when the lens is used for the observation instead of the unaided eye? Indicate the position of the lens and the eye with respect to the plane π, which provides for optimal observation of the fringes.

6.1.2.

A filter of optical density,

$$\Delta = 2 \qquad \left(\Delta = \log_{10} \frac{\text{Incident intensity}}{\text{Transmitted intensity}}\right)$$

is placed before the slit F_1. It is assumed that no phase change is introduced by the filter.

Determine the visibility V, of the fringes

$$V = \frac{I_{\max} - I_{\min}}{I_{\max} + I_{\min}}$$

* Problem 6 is by M. Rousseau and J. P. Mathieu, "Problèmes d'Optique," Dunod, Paris, 1966 (translated from the French by I. C. Gardner).

where I_{max} and I_{min} represent the maximum and minimum intensities, respectively.

6.1.3.

One uses a broad source illuminated by incoherent light.

6.1.3.1. The slit source has a height h (fixed) and a width a (variable). It is situated at a distance of 1 m behind the plane of the slits F_1 and F_2. Under these conditions what is the expression for the illumination at a point M in the plane π? How does the visibility of the fringes vary as function of a? From this law determine the phenomena observed as one progressively broadens the slit source F_0. If the lowering of the contrast is not to exceed 10%, determine the maximum width of the source.

6.1.3.2. To increase the luminosity of the image a grating of bright lines (parallel to F_1 and F_2) illuminated with incoherent light is used as a source. Determine the width, a, of the transparent lines and the spacing of the lines in order that the visibility retain the preceding value.

6.1.4.

6.1.4.1. One assumes that the slit source F_0 is sufficiently narrow to be considered as a line and one substitutes a photoelectric cell for the method of observation utilizing a Fresnel lens and the eye. The aperture of the photoelectric cell is taken to be a slit of fixed height and variable width b, oriented parallel to the fringes and located in the plane π. It is assumed that the intensity of the photoelectric current is proportional to the luminous flux falling upon the cell. How does the current vary as the cell and its aperture is moved perpendicular to the fringes (in the X direction)? Describe what happens as one progressively opens the slit.

6.1.4.2. What is the expression for the intensity of the photoelectric current when it is assumed that the source F_0 is not infinitely narrow but has the width a? determine the visibility factor.

6.1.5.

6.1.5.1. The width of the slit has the value of $a = 0.01$ mm and the width of the aperture of the photoelectric cell is $b = 0.02$ mm. Calculate the visibility factor.

The theoretical visibility factor V_t is greater than the experimental visibility factor V_r which has the value $V_r = 0.5$. Show that one can explain this result by assuming a parasitic signal $\mathscr{I}_0$ (background noise) which affects the cell in the absence of any luminous flux. Compute the ratio $\mathscr{I}_0/\mathscr{I}_{max}$ of the background noise to the maximum intensity of the signal.

6.1.5.2. The width of the aperture of the photoelectric cell is fixed at $b = 0.02$ mm by construction. On the other hand it is possible to modify the width of the slit source.

Compute V and show graphically its variation as a function of a. What value should one give a in order that V_r may be a maximum? What does one conclude from this study?

6.2. Solution

6.2.1. Coherent Illumination

6.2.1.1. Point Source. Designate by x and y the coordinates in the pupillary plane and let X and Y be the coordinates of a point M in the image plane (Fig. 6.1). The slits, infinitely narrow, diffract uniformly in the planes perpendicular to Oy.

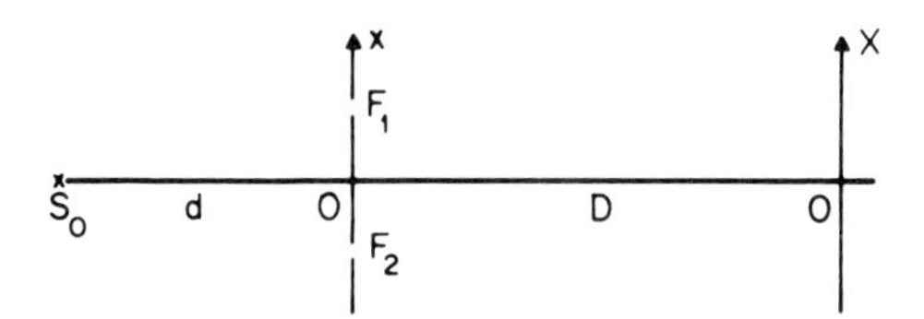

FIG. 6.1. Experimental set up.

The distribution of the illumination along the line OX is given by the equation

$$I = 4 \cos^2(\pi us) \tag{6.1}$$

where

$$u = \frac{\sin i}{\lambda} \approx \frac{i}{\lambda} = \frac{X}{D}\frac{1}{\lambda}. \tag{6.2}$$

This result can be obtained by noting that when the illumination is coherent, the distribution of the amplitude in the image is given by the Fourier transform (F.T.) of the distribution in the pupil.

Amplitude in the pupil plane

$$f(x) = \delta\left(x + \frac{s}{2}\right) + \delta\left(x - \frac{s}{2}\right). \tag{6.3}$$

Amplitude in the image plane

$$F(u) = \text{F.T. } [f(x)] \tag{6.4}$$

$$= \Delta(u)[e^{i\pi us} + e^{-i\pi us}] \tag{6.5}$$

with

$$\Delta(u) = \text{F.T. } [\delta(x)] = 1 \tag{6.6}$$

from which

$$F(u) = 2 \cos \pi us \qquad \left(\text{with period } \frac{2}{s}\right) \tag{6.7}$$

and

$$I(u) = |F(u)|^2 = 4 \cos^2 \pi us \qquad \left(\text{with period } \frac{1}{s}\right). \tag{6.8}$$

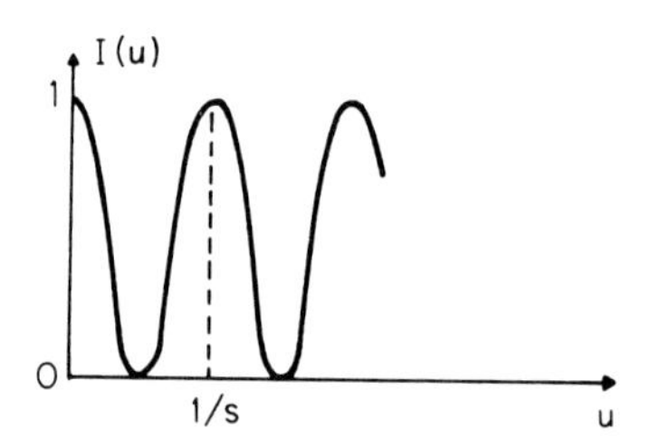

FIG. 6.2. Fringe distribution for a linear source.

6.2.1.2. Linear Source. There is no interference along lines parallel to Oy. Each point of the slit source produces a luminous line centered upon the geometric image and parallel to Ox. The fringes are parallel to F_1 and F_2 (Fig. 6.2).

The spacing of the fringes is given by the equation

$$\Delta u = \frac{1}{s} \tag{6.9}$$

and the linear distance between fringes, ΔX, is:

$$\Delta X = \lambda\left(\frac{D}{s}\right). \tag{6.10}$$

Numerical application:

$$\Delta X = 0.55 \times \frac{10^3}{2} = 0.275 \text{ mm.}$$

6.2.1.3. Observation of the Fringes.

6.2.1.3.1. UNAIDED EYE. An unaided eye at a distance of 25 cm has difficulty in resolving the image. The distance between two fringes subtens the angle

$$\varepsilon = \frac{0.275}{250} \approx 10^{-3} \text{ rad.}$$

This value is barely greater than the angular limit of resolution of the eye which is of the order of a minute or 3×10^{-4} rad.

6.2.1.3.2. LENS AND EYE. To avoid fatigue it is preferable that the eye be accommodated for infinity. To achieve this a magnifying lens (focal length 2 cm) is used with its focal plane coinciding with the plane π, projecting the image to an infinite distance. The image is then readily resolved because the angular separation of the fringes becomes

$$\varepsilon' = \frac{0.275}{20} = 0.0135 \text{ rad.}$$

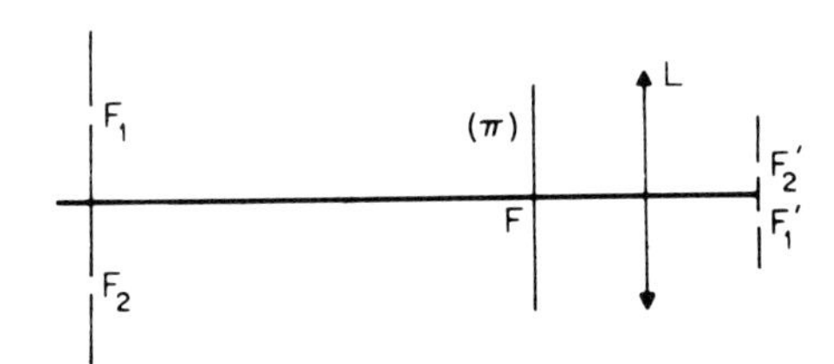

FIG. 6.3. Location of pupils.

The magnification of the lens is:

$$G = \frac{\varepsilon'}{\varepsilon} = \frac{\text{Angle subtended by the image}}{\text{Angle subtended for the unaided eye}}$$

$$G = 12.5.$$

Note: In principle the slits diffract through an angle of 180° and the lens, no matter how great the aperture, cannot collect all the rays. The observer, to receive the maximum flux, should place the pupil of his eye in the plane $F'_1F'_2$ conjugate to the plane F_1F_2 (Fig. 6.3).

The slits being at the distance ξ from the lens ($f = 2$ cm), the image will be at the distance ξ' such that

$$\frac{1}{\xi'} - \frac{1}{\xi} = \frac{1}{f}$$

$$\frac{1}{\xi'} = \frac{1}{100 + f} + \frac{1}{f} = \frac{52}{102}$$

$$\xi' = 1.965 \text{ cm} \approx 2 \text{ cm}.$$

The magnification is equal to:

$$\frac{\eta'}{\eta} = \frac{\xi'}{\xi} = \frac{1}{52}.$$

The dimension of the image is:

$$\eta' = \frac{\eta}{52} = \frac{2}{52} \approx 0.04 \text{ mm}.$$

All the rays incident on the lens will enter the eye provided that η' is less than the minimum diameter of the pupil of the eye.

6.2.2. Waves Passing through F_1 and F_2 in Phase which Differ in Amplitude

The intensity at the point M, where the difference in phase is φ, is given by

$$\begin{aligned} I(M) &= A_1^2 + A_2^2 + 2A_1A_2 \cos \varphi \\ &= I_1 + I_2 + 2(I_1I_2)^{1/2} \cos \varphi. \end{aligned} \tag{6.11}$$

The maximum and minimum intensities are given, respectively, by:

$$I_{\max} = (\sqrt{I_1} + \sqrt{I_2})^2$$
$$I_{\min} = (\sqrt{I_1} - \sqrt{I_2})^2.$$

The visibility factor is

$$V = \frac{2(I_1I_2)^{1/2}}{I_1 + I_2}. \tag{6.12}$$

Assume that a filter of optical density 2 is placed before the slit F_1:

$$\log_{10} \frac{I_2}{I_1} = 2 \qquad \text{or} \qquad \frac{I_2}{I_1} = 100$$

where $V = 0.2$ (see Fig. 6.4).

The position of the maxima and minima are the same with or without the filter. On the other hand, the contrast is unity only if the amplitudes are equal.

6.2.3. Broad Source, Incoherent Illumination

6.2.3.1. The Source as a Wide Slit. All points lying on a line parallel to Oy produce fringes parallel to Oy, the spacing corresponding to $\Delta u = 1/s$.

Consider the slit of width a to consist of an infinite number of infinitely narrow slits.

Let v be the reduced coordinate of a point in the plane of the source. The width of the source is given by:

$$v_0 = \frac{a}{\lambda d}. \tag{6.13}$$

The intensity at M produced by the element of width dv is

$$dI = A \times h\{1 + 2 \cos 2\pi[(u + v)s]\} dv. \tag{6.14}$$

A = const.

λvs = difference in path between the two waves arriving from F_1 and F_2.

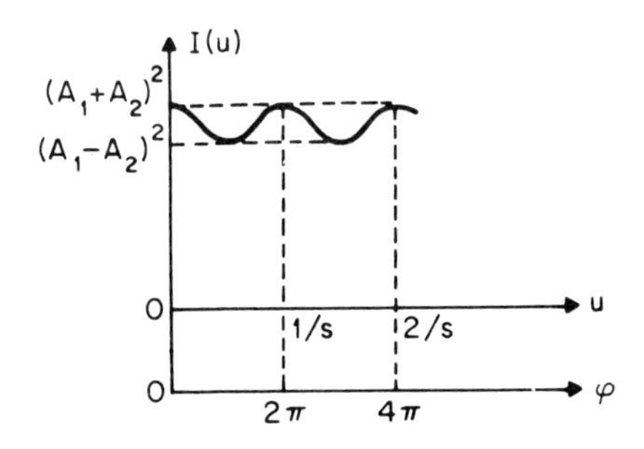

FIG. 6.4. Intensity distribution for out-of-phase amplitudes.

Each elementary slit of infinitesimal width gives a system of fringes of width defined by $\Delta u = 1/s$ and centered upon the geometric image of the elementary slit.

The intensity at M produced by the slit source is, therefore,

$$I = Ah \int_{-v_0/2}^{v_0/2} [1 + \cos 2\pi(u + v)s] \, dv \tag{6.15}$$

$$= I_0 \left[1 + \frac{\sin \pi v_0 s}{\pi v_0 s} \cos 2\pi us \right]. \tag{6.16}$$

From this it can be derived that

$$V = \frac{\sin \pi v_0 s}{\pi v_0 s}.$$

The variation of V with v_0 is shown in Fig. 6.5.

6.2.3.1.1. Numerical Application. V will be less than, or equal to, 0.9 if

$$\frac{\sin \pi v_0 s}{\pi v_0 s} = 0.9 \qquad \text{or} \qquad \pi v_0 s = \frac{\pi}{4} \qquad \text{or} \qquad v_0 = \frac{1}{4s}.$$

From the definition of v_0

$$\frac{1}{4s} = \frac{a}{\lambda d} \qquad \text{where} \quad a = d\frac{\lambda}{4s} = 10^6 \times \frac{0.55 \times 10^{-3}}{4 \times 2}$$

$$V = 0.9 \qquad \text{for} \quad a \approx 70\ \mu\text{m}.$$

The fringes disappear for $a = 275\ \mu$m.

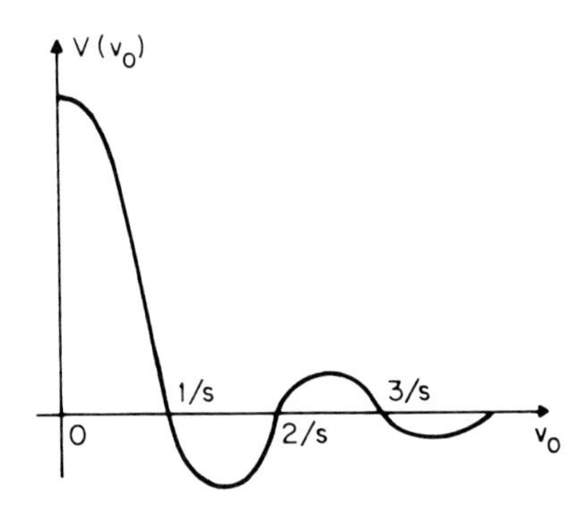

Fig. 6.5. Visibility for a broad source.

The theorem of Van Cittert–Zernicke permits this result to be obtained immediately. The degree of coherence between the slits F_1 and F_2 is given by the Fourier transform of the distribution of intensity in the plane of the source. Since the problem is one-dimensional it will be sufficient to consider the source as a slit of width a, parallel to Oy and that the pupil consists of two apertures P_1 and P_2 of negligible diameter, separated by the distance s, in an opaque screen (P_1 and P_2 correspond to the intersections of the two slits F_1 and F_2 with the line Ox). The distribution of the intensity in the source is represented by the rectangular function shown in Fig. 6.6.

$$I(v) = \begin{cases} 0 & \text{for} \quad |v| < -v_0/2 \text{ and } |v| > +v_0/2, \\ 1 & \text{for} \quad -v_0/2 < |v| < +v_0/2. \end{cases} \tag{6.17}$$

From this:

$$\text{F.T. } [I(v)] = \varphi(x) = \frac{\sin \pi v_0 x}{\pi v_0 x} \tag{6.18}$$

(and Figs. 6.6 and 6.7).

One now places the assumed diffracting area to bring its center into coincidence with P_1 (Fig. 6.7). The visibility of the fringes is equal to the value of $\varphi(x)$ for the abscissa corresponding to P_2, that is to say $\varphi(s)$ (Fig. 6.7). It is evident that the contrast of the fringes will be good for $s = \frac{1}{4}v_0$.

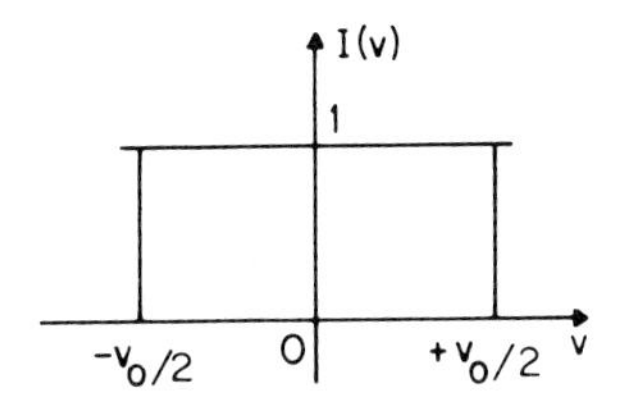

FIG. 6.6. Intensity distribution at the source.

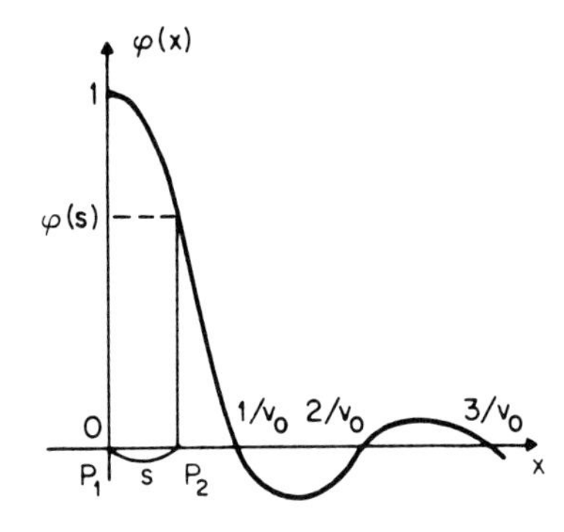

FIG. 6.7. Visibility of the fringes.

6.2.3.2. The Source—A Grating Illuminated by Incoherent Light. The reduced coordinate corresponding to the interval p of the grating will be designated v_p.

(a) It will first be assumed that the illuminated lines of the grating are infinitely narrow.

The distribution of intensity in the source is a "Dirac comb" (Fig. 6.8). Its Fourier transform is a "Dirac comb" with the interval between successive maxima equal to $1/v_p$ (Fig. 6.9).

As before, place the assumed diffracting area $\varphi(x)$ upon the pupil so that $\varphi(0)$ coincides with the point P_1.

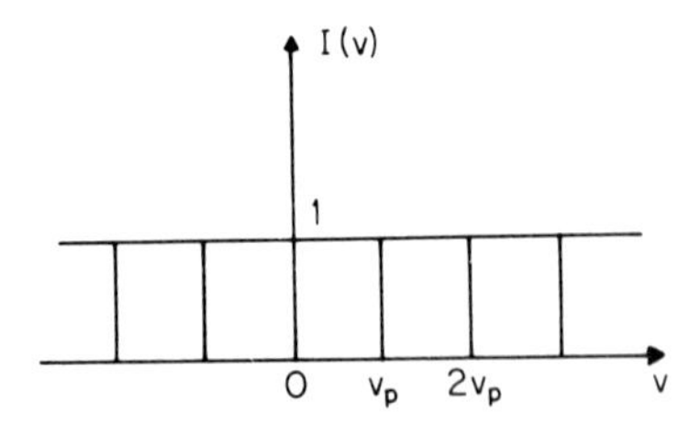

FIG. 6.8. "Dirac-comb" distribution at the source.

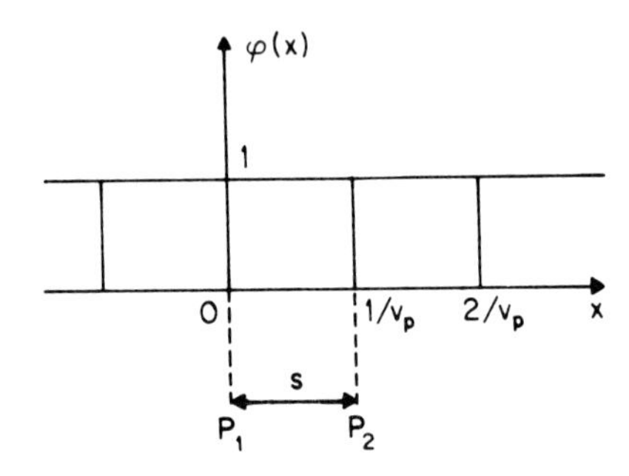

FIG. 6.9. Fringe contrast for "Dirac-comb" source.

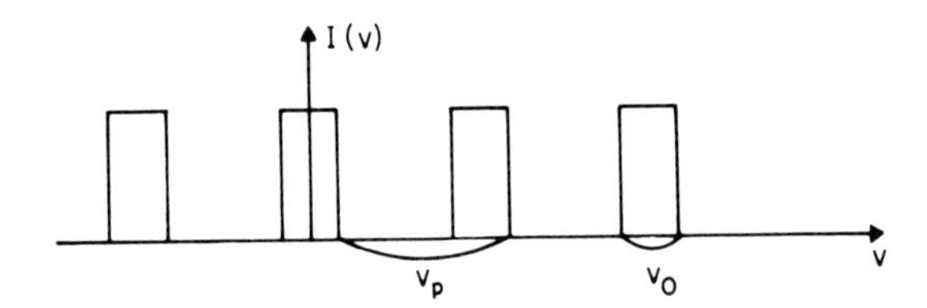

FIG. 6.10. Square wave intensity distribution of the source.

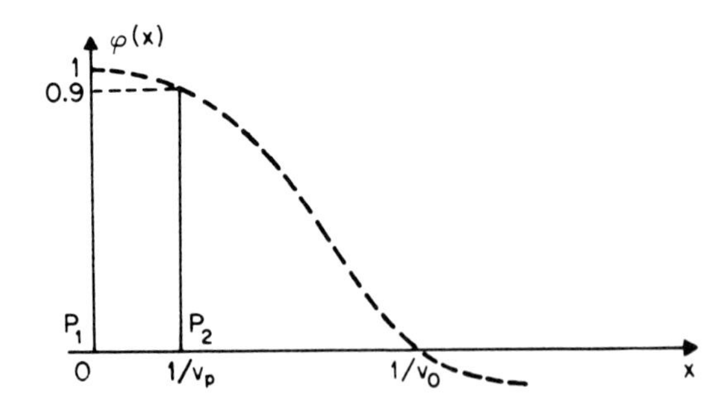

FIG. 6.11. Fourier transform of the intensity given in Fig. 6.10.

The contrast of the fringes will be one (Fig. 6.9) if

$$\frac{1}{v_p} = s.$$

That is to say

$$s = \frac{\lambda d}{p}$$

and

$$p = \frac{\lambda d}{s} = 0.55 \times \frac{10^3}{2} = 275\ \mu\text{m}.$$

(b) The lines of the grating will be assumed to have the finite width a. $I(v)$ is a square wave as represented in Fig. 6.10, the spacing equal to v_p and the width of each area equal to v_0. The Fourier transform of $I(v)$ is represented by Fig. 6.11. In order that the image (interference pattern) may have acceptable contrast it is necessary that

$$s = \frac{1}{v_p} = \frac{1}{4v_0}.$$

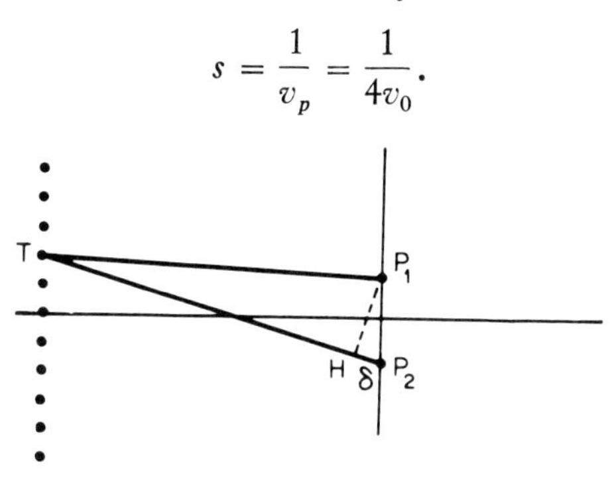

FIG. 6.12. Simplified way of estimating the intensity distribution.

6.2.3.2.1. NUMERICAL RESULTS. p, the spacing between successive lines $= 275\ \mu$m. The width of an illuminated line $a = 70\ \mu$m.

Note: One is able to obtain the solution by a very simple consideration (Fig. 6.12).

(a) Grating with infinitely narrow lines: The fringes remain unchanged if the phase difference between the waves from any two lines is an integral multiple 2π when they arrive at P_1 and P_2.

(b) Grating with wide lines: The waves proceeding from the edge of a slit shall present at P_1 and P_2 a path difference lying between $k\lambda$ and $(k + \frac{1}{8})\lambda$ in order that the fringes may not be washed out. (The fringes produced by the extreme edges of a slit will be displaced not more than one-quarter of the distance between fringes.)

6.2.4. Detector Aperture — a Finite Width

6.2.4.1. The Slit Source—Infinitely Narrow. The fringes on the plane π have unit contrast (see question I). On the other hand, because of the finite width of the aperture of the sensor, the flux received by the sensor is never zero (Fig. 6.13). The illumination is the same for all points lying in a vertical line in the plane of observation. It is assumed that the aperture of the sensor is comprised of elements of width du and of height l.

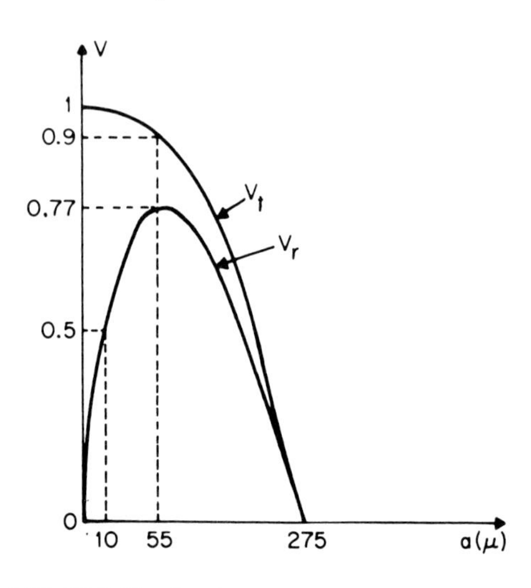

FIG. 6.13. Visibility for detector aperture with finite width.

Let u_c be the reduced coordinate corresponding to the linear width b of the aperture of the sensor. The flux which is received by an element of surface du of abscissa u' is

$$d\Phi = Bl(1 + \cos 2\pi us)\, du \tag{6.19}$$

where

$$\Phi(u') = \int_{u'-u_c/2}^{u'+u_c/2} d\Phi = Blu_c\left[1 + \frac{\sin \pi u_c s}{\pi u_c s} \cos 2\pi u' s\right]. \tag{6.20}$$

As earlier, the coefficient of visibility may be defined as:

$$V = \frac{\sin \pi u_c s}{\pi u_c s}. \tag{6.21}$$

So long as u_c is less than $\frac{1}{4}s$, the intensity of the photoelectric current, proportional to the luminous flux, is essentially sinusoidal. As one opens the aperture the difference between the maximum and minimum values becomes less. Finally, when the aperture is opened to the width $u_c = 1/s$, the intensity of the current does not vary as one moves the sensor.

6.2.4.2. The Slit Source —— Finite Width a. One has

$$I(u) = I_0\left[1 + \frac{\sin \pi v_0 s}{\pi v_0 s} \cos 2\pi us\right] \tag{6.22}$$

whence

$$\Phi'(u') = Blv_0 \int_{u'-u_c/2}^{u'+u_c/2} \left[1 + \frac{\sin \pi v_0 s}{\pi v_0 s} \cos 2\pi us\right] du \tag{6.23}$$

$$\Phi(u') = Blu_c v_0 \left[1 + \frac{\sin \pi u_c s}{\pi u_c s} \times \frac{\sin \pi v_0 s}{\pi v_0 s} \cos 2\pi u's\right]. \tag{6.24}$$

From this one can derive:

$$V = \frac{\sin \pi u_c s}{\pi u_c s} \times \frac{\sin \pi v_0 s}{\pi v_0 s}. \tag{6.25}$$

The coefficient of visibility which is a "function of the apparatus" may be defined as

$$V = \text{F.T. (slit source width } v_0) \times \text{F.T. (slit cell width } u_c).$$

In the special case, when the aperture of the source is infinitely narrow, the first term of the product reduces to one since the Fourier transform of $\delta(x) = 1$.

6.2.5. Influence of Background Noise

Return to Eq. (6.25) giving the coefficient of theoretical visibility.

(1) $a = 0.01$ mm, $b = 0.02$ mm ($v_0 = \text{const}$, $u_c = \text{const}$).

One has

$$V_t = 0.991 \times \frac{\sin \pi v_0 s}{\pi v_0 s} = 0.987.$$

Taking into account the background noise, the actual value of the current from the sensor is

$$\mathscr{I}_r(u) = \mathscr{I}(u) + \mathscr{I}_0, \tag{6.26}$$

from which the coefficient of experimental visibility is

$$V_r = \frac{\mathscr{I}_{r\max} - \mathscr{I}_{r\min}}{\mathscr{I}_{r\max} + \mathscr{I}_{r\min}} = \frac{\mathscr{I}_{\max} - \mathscr{I}_{\min}}{\mathscr{I}_{\max} + \mathscr{I}_{\min} + 2\mathscr{I}_0}. \tag{6.27}$$

$$V_r = \frac{(\sin \pi u_c s/\pi u_c s)\,(\sin \pi v_0 s/\pi v_0 s)}{1 + \mathscr{I}_0/v_0}. \tag{6.28}$$

It has been assumed that the constant coefficient Blu_c is equal to one. One therefore has the relation:

$$V_r = \frac{V_t}{1 + \mathscr{I}_0/v_0}. \tag{6.29}$$

6.2.5.1. Numerical Application

$$1 + \frac{\mathscr{I}_0}{v_0} = \frac{V_t}{V_r} = \frac{0.987}{0.5} = 1.974.$$

$$\frac{\mathscr{I}_0}{v_0} = 0.974.$$

One has

$$\mathscr{I}_{\max} = v_0[1 + V_t] = v_0[1.987]$$

where

$$\frac{\mathscr{I}_0}{\mathscr{I}_{\max}} = \frac{0.974}{1.987} \approx \frac{1}{2}$$

(in reality this value is smaller).

(2) a variable, $b = 0.02$ mm (v_0 variable, $u_c = \text{const}$).

One has

$$V_t = 0.991\,\frac{\sin \pi\, v_0 s}{\pi v_0 s} \tag{6.30}$$

whence

$$V_r = \frac{0.991\,(\sin \pi\, v_0 s/\pi v_0 s)}{1 + \mathscr{I}_0/v_0} = 0.991 \times \pi s\,\frac{\sin \pi\, v_0 s}{\pi s(v_0 + \mathscr{I}_0)}.$$

V_r passes through a maximum for $dV_r/dv_0 = 0$

$$\tan \pi\, v_0 s = \pi v_0 s + \pi s \mathscr{I}_0 = \pi v_0 s + 0.111.$$

This relation is verified for $\pi v_0 s \approx 35°$

$$v_0 = \frac{36}{180}\,\frac{1}{2 \times 10^3} = 10^{-4}\ \mu\text{m}^{-1} \qquad \text{or} \qquad v_0 = \frac{a}{\lambda f},$$

whence

$$a = v_0 f\lambda = 10^{-4} \times 10^6 \times 0.55$$

$$a = 0.55\ \mu\text{m}.$$

In Fig. 6.13 the variations of V_t and V_r are shown as a function of v_0.
For $a = 55$ μm one has

$$V_t = 0.991 \times \frac{\sin 35°}{(3.14 \times 36)/180} = 0.89$$

whence

$$V_r = \frac{V_t}{1 + \mathscr{I}_0/v_0} = \frac{0.9}{1 + 0.974 \times \frac{10}{55}} = \frac{0.9}{1.177}$$

$$V_{r\,\max} = 0.77.$$

6.3. Conclusion

In theory, to have the most favorable contrast the slit source should be narrowed as much as possible. In practice, if there is much background noise, it is necessary to widen the slit source to the optimal width.

MAGNETIC FIELD CALCULATION OF AXIAL SYMMETRIC SYSTEMS (PROBLEMS 7–14)†

7. FIELD OF LINE CURRENTS*

7.1. Problem

Determine the magnetic field induced by line currents by applying Biot Savart and Ampère laws established experimentally, relating the magnetic field **B** to the current **I**, and postulate the law of force between line currents.

7.2. Solution

The Biot–Savart law is expressed in vector notation

$$d\mathbf{B} = \frac{\mu_0}{4\pi} I \frac{d\mathbf{l} \times \mathbf{r}}{|\mathbf{r}|^3} \tag{7.1}$$

which can be rewritten in integral form:

$$\mathbf{B} = \frac{\mu_0 I}{4\pi} \operatorname{curl} \int_c \frac{d\mathbf{l}}{|\mathbf{r}|}. \tag{7.2}$$

The magnetic vector potential **A** is derived from Eq. (7.2):

$$\mathbf{A} = \frac{\mu_0}{4\pi} I \int \frac{d\mathbf{l}}{|\mathbf{r}|}. \tag{7.3}$$

Equation (7.3) is a convenient way to calculate **A** and **B** at a point P due to line currents flowing through a filamentary wire C, as illustrated in Fig. 7.1.

If the total current occupies a volume v and passes through a surface S, bound by the contour C, the vector potential is expressed by:

$$\mathbf{A} = \frac{\mu_0}{4\pi} \int dI \int \frac{d\mathbf{l}}{|\mathbf{r}|}. \tag{7.4}$$

† A list of suggested references for Problems 7–14 can be found at the end of Problem 14.

* Problem 7 is by H. Brechna.

We link the current I to the current density by the relation

$$dI \, dl = \mathbf{J} \, dv$$

and we get from Eqs. (7.2) and (7.4)

$$\mathbf{A} = \frac{\mu_0}{4\pi} \int_v \frac{\mathbf{J}}{|\mathbf{r}|} \, dv \tag{7.5}$$

$$\mathbf{B} = \frac{\mu_0}{4\pi} \int_v \left[\mathbf{J} \cdot \frac{\mathbf{r}}{|\mathbf{r}|^3} \right] dv. \tag{7.6}$$

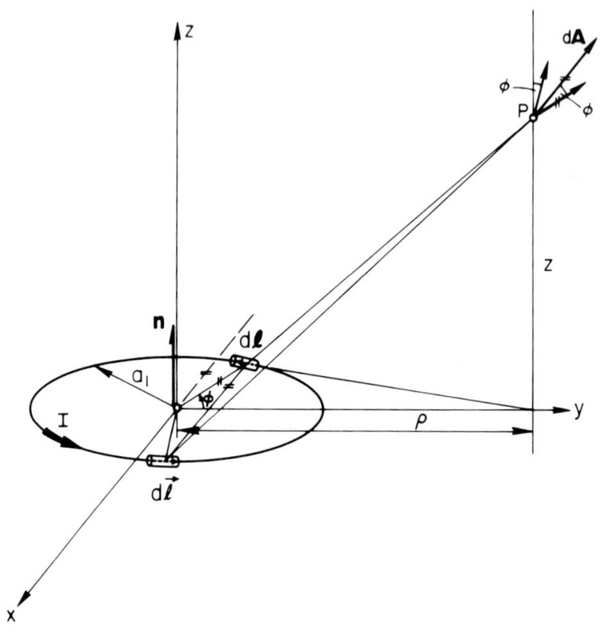

FIG. 7.1.

If the observation point P is outside the surface S, as in the case of Fig. 7.2, then $\nabla(1/r) = 0$. We define

$$\Omega = -\int \mathbf{n} \operatorname{grad} \left(\frac{1}{r} \right) ds = \int_s \frac{\cos \Theta}{r^2} \, ds \tag{7.7}$$

and call Ω the solid angle. Combining Eqs. (7.7) and (7.6), we may derive the relation between B, I, and Ω:

$$\mathbf{B} = -\frac{\mu_0 I}{4\pi} \, \mathbf{grad} \, \Omega. \tag{7.8}$$

If the surface of integration *S includes* the observation point *P*, we express the solid angle as follows:

$$S = r^2 \, d\Omega = r^2 \sin \Theta \, d\Theta \, d\Phi.$$

Thus the solid angle is given by:

$$d\Omega = \sin \Theta \, d\Theta \, d\Phi. \tag{7.9}$$

Integrating (7.9) over the sphere of radius unity, we get:

$$\Omega = 4\pi.$$

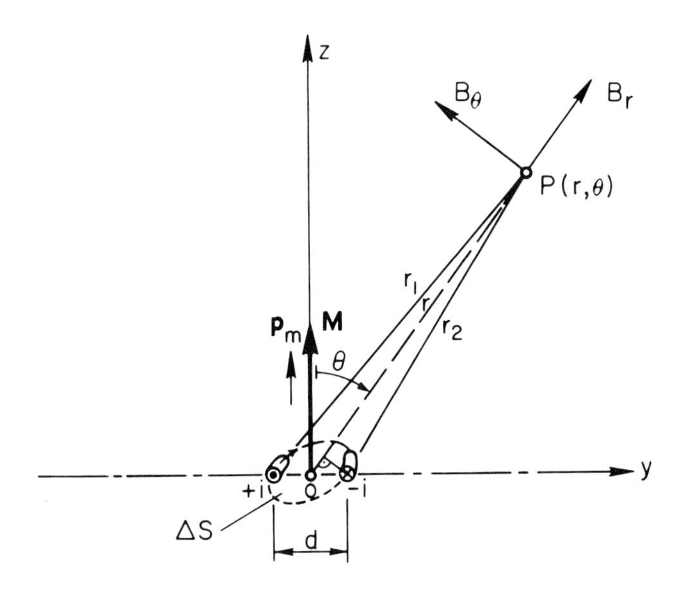

FIG. 7.2.

Magnetomotive force between points P_1 and P_2 is given by

$$\text{MMF} = \int_{P_2}^{P_1} \mathbf{H} \, d\mathbf{r}. \tag{7.10}$$

For a circular path about a current (I) carrying wire located at the center of the path, the field intensity is

$$\mathbf{H} = \frac{I}{2\pi r} \cdot \hat{\Theta} \tag{7.11}$$

which yields the MMF:

$$\text{MMF} = \oint \mathbf{H}\, dr = I \qquad \text{(amperes)}. \tag{7.12}$$

In the case of a closely wound toroidal coil,

$$\text{MMF} = NI \tag{7.13}$$

with N as the number of turns.

The force on a conductor carrying a current I is

$$F = \int \mathbf{I}\, dl \cdot \mathbf{B}. \tag{7.14}$$

If two conductors located at a distance r carry the currents I_1 and I_2, the force per unit length of conductor 2 is:

$$\boldsymbol{F} = \frac{\mu_0}{2\pi} \cdot \frac{I_1 \cdot I_2}{r} \cdot l_2 \cdot \hat{r}. \tag{7.15}$$

7.3. Examples

7.3.1.

The components of field intensity of a circular loop with radius a, carrying a current I, at a point P on the axis of symmetry, are

$$H_r = 0 \tag{7.16}$$

$$H_z = \frac{Ia^2}{2(a^2 + z^2)^{3/2}}\,. \tag{7.17}$$

7.3.2.

The field intensity inside a conductor with radius R, carrying a current I, is given by

$$H = \frac{Ir}{2\pi R^2} \tag{7.18}$$

with r the distance of a point inside the conductor from the wire center.

7.3.3.

The field intensity of an infinitely long solenoid at any point on the axis is

$$H = NI \tag{7.19}$$

independent of the radius of the solenoid. The factor N is the number of turns per unit axial length.

7.3.4.

The force per unit length of two straight and parallel conductors located at a distance d from each other, carrying the currents I_1 and I_2, is

$$\frac{F}{l} = \frac{\mu_0}{2\pi}\frac{I_1 \ I_2}{d}. \qquad (7.20)$$

7.3.5.

The force between two coaxial solenoids with their winding planes perpendicular to the axis of symmetry, carrying the currents I_1 and I_2, and having N_1 and N_2 turns, is given by

$$F = \frac{\mu_0}{2\pi}\frac{N_1 I_1 \cdot N_2 I_2}{d} 2\pi a \qquad (7.21)$$

with d the distance of central filaments of radius a.

7.3.6.

The magnetic vector potential and the magnetic field intensity in the x, y plane of a straight conductor of length $2s$ is given by

$$A_y = \frac{I}{4\pi}\left[\ln[s + (x^2 + s^2)^{1/2}] - \ln x\right] \qquad (7.22)$$

$$H_z = -\frac{I}{2\pi x}. \qquad (7.23)$$

7.3.7.

The magnetic field established by a current I flowing through two long parallel conductors located in the x, y plane is

$$H_x = -\frac{I}{2\pi}\left(\frac{z}{r_1^2} - \frac{z}{r_2^2}\right) \qquad (7.24)$$

where

$$r_1 = \left[\left(x + \frac{s}{2}\right)^2 + x^2\right]^{1/2}, \quad r_2 = \left[\left(x - \frac{s}{2}\right)^2 + x^2\right]^{1/2}.$$

To obtain Eq. (7.24), use the magnetic vector potential

$$A_y = \frac{I}{2\pi}\left(\ln r_2 - \ln r_1\right).$$

8. FIELD OF A CIRCULAR LOOP*

8.1. Problem

Calculate the magnetic flux density of a circular loop, using expressions derived for the solid angle.

8.2. Solution

We place the observation point on the symmetry axis of a circular line current of radius a (Fig. 8.1). The solid angle is expressed by:

$$\Omega = \int_{\alpha}^{0} \int_{0}^{2\pi} \sin\alpha \, d\alpha \, d\Phi = 2\pi \int_{\cos\alpha}^{0} d(\cos\alpha)$$

$$= 2\pi(1 - \cos\alpha). \tag{8.1}$$

The scalar potential of the circular line current is derived from

$$V = \frac{I}{4\pi} \Omega \tag{8.2}$$

which yields

$$V = \frac{I}{2}(1 - \cos\alpha) = \frac{I}{2}\left(1 - \frac{z}{r}\right) \tag{8.3}$$

with

$$r = (z^2 + a^2)^{1/2}$$

$$\frac{dr}{dz} = \frac{z}{r}.$$

The axial component of the magnetic field is expressed by

$$B_z = -\mu_0 \frac{dV}{dz} = \frac{\mu_0 I}{2} \frac{a^2}{r^3}$$

$$= \mu_0 \frac{I}{2a} \sin^3 x. \tag{8.4}$$

* Problem 8 is by H. Brechna.

If the observation point is placed in the center of the circle, i.e., $r \to a$, the axial field has the form:

$$B_0 = \mu_0 \frac{I}{2a}. \tag{8.5}$$

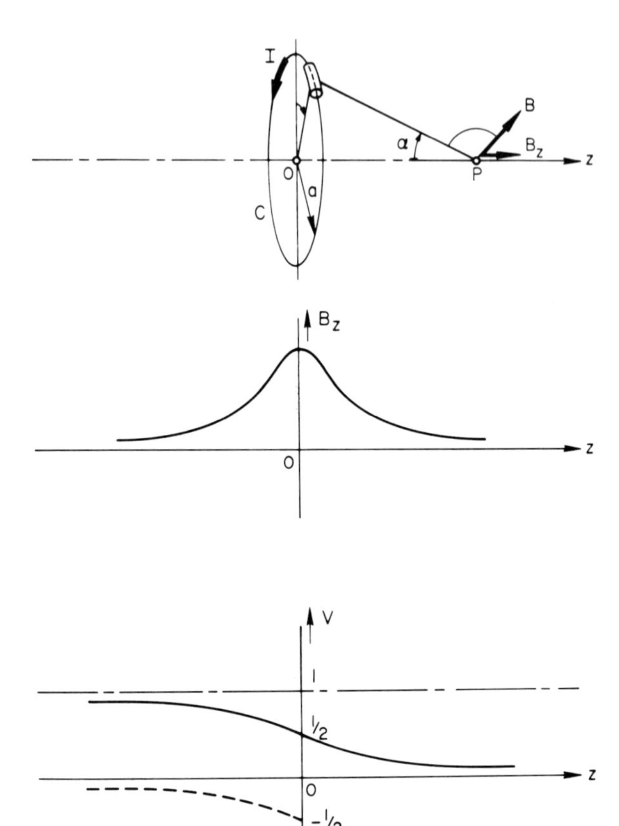

FIG. 8.1.

9. DIPOLE FIELDS*

9.1. Problem

Calculate the magnetic field by means of distributed volume dipoles.

9.2. Solution

Field calculation by means of magnetic volume dipoles is useful not only in determining the field contribution of ferromagnetic materials, but also to determine the fields of line and surface currents. Assuming a magnetic

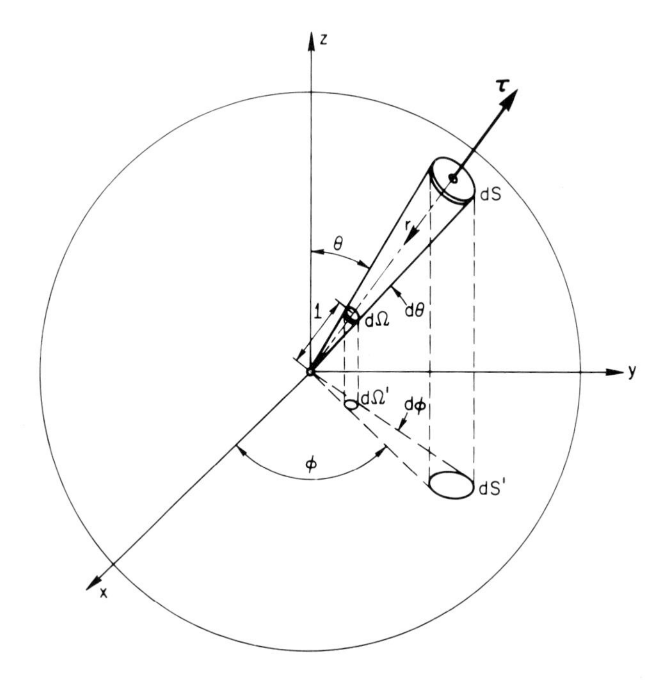

FIG. 9.1.

* Problem 9 is by H. Brechna.

dipole is oriented at an angle Θ to the z axis (Fig. 9.1), the potential of the dipole is given by:

$$V_m = \frac{\mathbf{M}}{4\pi} \frac{\cos \Theta}{r^2}. \tag{9.1}$$

We express V_m in vector notation

$$V_m = -\frac{1}{4\pi}\left[\mathbf{p}_m \cdot \mathbf{grad}\, \frac{1}{r}\right], \tag{9.2}$$

where $\mathbf{p}_m$ is the magnetic moment and is expressed by

$$\mathbf{p}_m = \int \mathbf{M}\, dv. \tag{9.3}$$

The flux density can be derived from Eq. (9.2) in the form

$$\mathbf{B}_m = \frac{1}{4\pi}\left[\frac{3}{r^5}(\mathbf{p}_m \cdot \mathbf{r})\,\mathbf{r} - \frac{\mathbf{p}_m}{r^3}\right], \tag{9.4}$$

or in terms of the dipole moment

$$\mathbf{B} = \frac{1}{4\pi} \frac{3\hat{n}\,(\mathbf{M} \cdot \hat{n}) - \mathbf{M}}{r^3}, \tag{9.5}$$

where $\hat{n}$ denotes the unit vector along $\mathbf{r}$, and $\mathbf{M}$ the magnetization.

10. FIELDS DUE TO MAGNETIZED DIPOLES*

10.1. Problem

Find the field components B_z and B_r in the z and r directions due to the magnetization **M** in polar coordinates (Fig. 10.1).

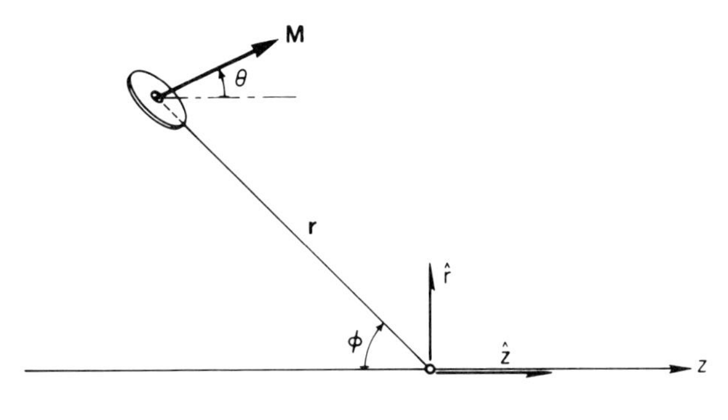

FIG. 10.1.

10.2. Solution

If $\hat{z}$ and $\hat{r}$ denote the unit vectors in z and r directions, we get the axial field component:

$$\begin{aligned} B_z &= B \cdot \hat{z} \\ &= \frac{3 \cdot (\hat{n} \cdot \hat{z})\,(\mathbf{M} \cdot \hat{n}) - \mathbf{M} \cdot \hat{z}}{4\pi \,|\, r \,|^{3}}, \end{aligned} \tag{10.1}$$

and the radial field component:

$$\begin{aligned} B_r &= B \cdot \hat{r} \\ &= \frac{3\,(\hat{n} \cdot \hat{r})\,(M \cdot \hat{n}) - M \cdot}{4\pi \,|\, r \,|^{3}}. \end{aligned} \tag{10.2}$$

* Problem 10 is by H. Brechna.

Expressing Eqs. (10.1) and (10.2) in terms of the angles Θ and Φ, we may write:

$$B_z = \frac{3M \cos \Phi \cos(\Phi + \Theta) - M \cos \Theta}{4\pi \mid r \mid^3} \tag{10.3}$$

$$B_r = \frac{-3M \sin \Phi \cos(\Phi + \Theta) - M \sin \Theta}{4\pi \mid r \mid^3}. \tag{10.4}$$

11. MAGNETIC FIELDS OF A CURRENT LOOP*

11.1. Problem

If we desire to calculate the field components of current loops of radius a, several possibilities are explored. We treat each case separately.

11.2. Solution Using the Law of Biot–Savart

The magnetic vector potential was given in the form:

$$\mathbf{A} = \frac{\mu_0 I}{4\pi} \int_c \frac{dl}{r}$$

which by referring to Fig. 7.1 yields:

$$\mathbf{A} = \frac{\mu_0}{4\pi} I \int_{\Phi=0}^{2\pi} \frac{\hat{m} \cdot a \cdot \sin \Phi \, d\Phi + \hat{n} \cdot a \cdot \cos \Phi \, d\Phi}{[(\rho - a \cos \Phi)^2 + (a \sin \Phi)^2 + z^2]^{1/2}} \cdot \tag{11.1}$$

Due to the condition of axial symmetry, only the Φ component of $\mathbf{A}$ exists,

$$A_\Phi = \frac{\mu_0 I}{2\pi} \int_0^\pi \frac{a \cos \Phi \, d\Phi}{[(\rho - a \cos \Phi)^2 + z^2]^{1/2}} \tag{11.2}$$

which, integrated, yields:

$$A_\Phi = \frac{\mu_0 I k}{2\pi} \left(\frac{a}{\rho}\right)^{1/2} \left\{\left(\frac{2}{k^2} - 1\right) K(k) - \frac{2}{k^2} E(k)\right\} \tag{11.3}$$

with:

$$K(k) = \int_0^{\pi/2} \frac{d\Psi}{(1 - k^2 \sin^2 \Psi)^{1/2}}$$

$$E(k) = \int_0^{\pi/2} (1 - k^2 \sin^2 \psi)^{1/2} \, d\Psi$$

as the complete first and second kinds of elliptic integrals.

* Problem 11 is by H. Brechna.

In deriving Eq. (11.3) we used the notations:

$$k^2 = \frac{4a\rho}{(a + \rho)^2 + z^2}$$

$$\rho^2 = x^2 + y^2$$

$$\Phi = \pi + 2\Psi.$$

The components of the magnetic flux density at the observation point P are given in cylindrical coordinates:

$$B_\rho = -\frac{\partial A_\Phi}{\partial z}$$

$$= \frac{\mu_0}{2\pi} I \frac{z}{\rho} [(\rho + a)^2 + z^2]^{-1/2} \left[-K(k) + \frac{a^2 + \Phi^2 + z^2}{(a - \rho)^2 + z^2} E(k) \right] \tag{11.4}$$

$$B_\Phi = 0$$

$$B_z = \frac{1}{\rho} \frac{\partial}{\partial \rho} (\rho A_\rho)$$

$$= \frac{\mu_0}{2\pi} I [(\rho + a)^2 + z^2]^{-1/2} \left[K(k) + \frac{a^2 - \rho^2 - z^2}{(a - \rho)^2 + z^2} E(k) \right]. \tag{11.5}$$

Along the symmetry axis z where $\rho = 0$ we get $k = 0$, $K(0) = E(0) = \pi/2$, and thus:

$$B_\rho = 0 \tag{11.6}$$

$$B_z = \frac{\mu_0 I}{2} \frac{\sin^2 \Theta}{r}. \tag{11.7}$$

At the center of the loop with $z = 0$ (for $a = a_1$), $B_z = (\mu_0/2a_1) I$. If a constant current I flows through a loop of radius a_1, the field components can be calculated from (11.4) and (11.5) at any point in space. For constant values of a and I, values of $(4\pi a_1/\mu_0 I) B_\rho$ and $(4\pi a_1/\mu_0 I) B_z$ are illustrated as a function of ρ/a_1 in Fig. 11.1 and 11.2.

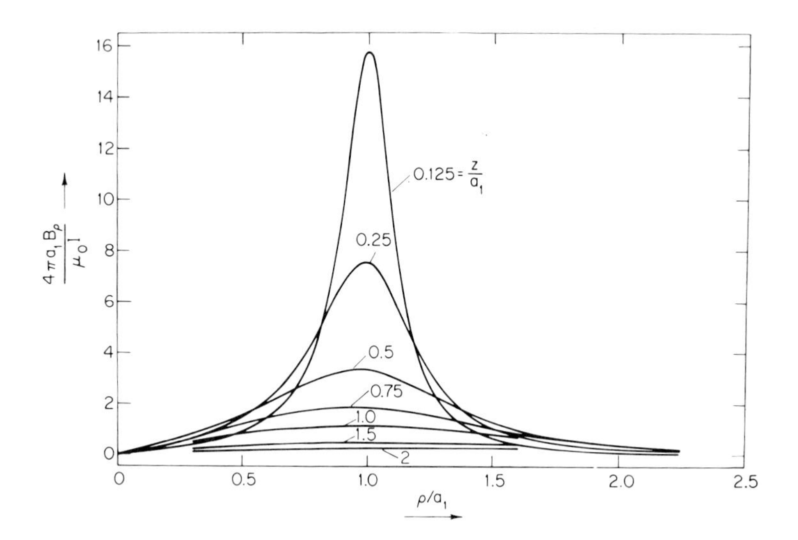

FIG. 11.1.

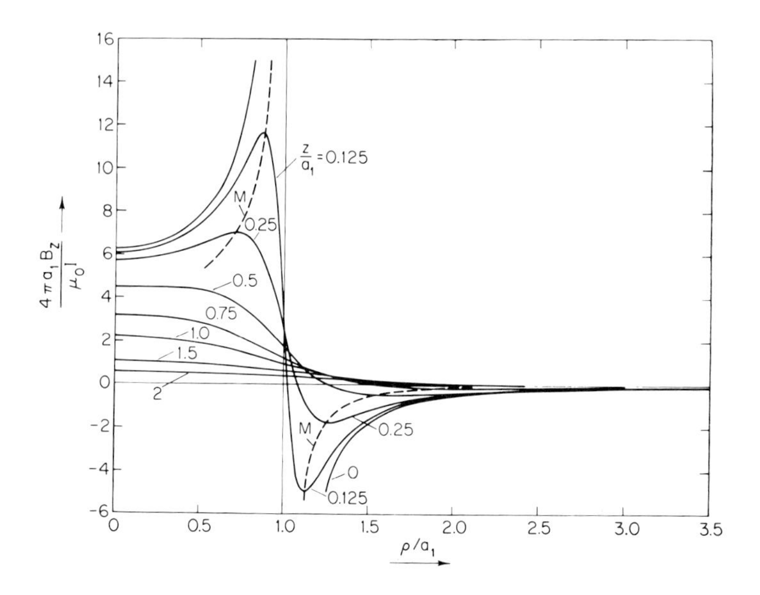

FIG. 11.2.

11.3. Find the Scalar Potential V at a Point $P(\rho, z)$ Generated by a Current Loop of Radius a Assuming the Current I Flows through the Loop

If the potential on the axis of rotation is known, the potential at any point in space may be derived by using Taylor's expansion:

$$V(\rho, z) = V(0, z) - \frac{\rho^2}{4}\frac{d^2}{dz^2} V(0, z) + \frac{\rho^4}{64}\frac{d^4}{dz^4} V(0, z) \cdots - \frac{(-1)^n}{(n!)^2}\left(\frac{\rho}{2}\right)^{2n}\left(\frac{d}{dz}\right)^{2n} V(0, z). \quad (11.8)$$

In Problem 8 we calculated the potential of a current loop along the axis of symmetry z:

$$V(0, z) = \frac{I}{2}\left(1 - \frac{z}{r}\right) = \frac{I}{2}\left(1 - \frac{\partial r}{\partial z}\right) \quad (11.9)$$

where $r = (a^2 + z^2)^{1/2}$. Differentiating $V(0, z)$ with respect to z yields

$$\frac{d^2V(0, z)}{dz^2} = -\frac{I}{2}\frac{\partial^3 r}{\partial z^2} = 1.3\frac{I}{2r^5}(a^2 z)$$

$$\frac{d^4V}{dz^4}(0, z) = -\frac{I}{2}\frac{\partial^5 r}{\partial z^5} = -3.5\frac{I}{2r^9}a^2 z[3a^2 - 4z^2]$$

$$\frac{d^6V}{dz^6}(0, z) = -\frac{I}{2}\frac{\partial^7 r}{\partial z^7} = +3\cdot 5\cdot 7\frac{I}{2r^{13}}a^2 z[15a^4 - 64a^2z^2 + 24z^4]$$

$$\vdots \quad (11.10)$$

$$\frac{d^{2m}V}{dz^{2m}}(0, z) = -\frac{I}{2}\frac{\partial^{2m+1}}{\partial z^{2m+1}}V(0, z)$$

$$= \frac{(-1)^{m+1}\,1\cdot 3\cdot 5\cdots(2m+1)\,I}{2r^{4m+1}} \times a^2 z\,[c_{1,m}\,a^{2m-2} - c_{2,m}\,a^{2m-4}z^2 + c_{3,m}\,a^{2m-6}z^4 - \cdots]$$

where

$$c_1 = 1\cdot 3\cdot 5\cdots(2m-1)$$

$$c_{m,m} = 1(2+2)(3+3)(\cdots)(m+m).$$

11.3.1. Example

Calculate the magnetic field at a point z on the axis of symmetry using (11.8) and compare the result with Eqs. (7.17), (8.4), and (11.5).

11.4. Calculate Magnetic Field Intensity of a Current Loop of Radius a at a Point $P(\rho, z)$, Using the Magnetic Scalar Potential Expansion of Section 11.3

By using the relation $\mathbf{H} = -\,\mathbf{grad}\, V$, the components of H in cylindrical coordinates are given by:

$$H_z(\rho, z) = -\frac{\partial V}{\partial z} = H(0, z) - \frac{\rho^2}{4}\left(\frac{d}{dz}\right)^2 H(0, z) + \frac{\rho^4}{64}\left(\frac{d}{dz}\right)^4 H(0, z)$$

$$\mp \cdots \frac{(-1)^n}{(n!)^2}\left(\frac{\rho}{2}\right)^{2n}\left(\frac{d}{dz}\right)^{2n} H(0, z) \cdots \tag{11.11}$$

$$H_\rho(\rho, z) = -\frac{\rho}{2}\frac{d}{dz} H(0, z) + \frac{\rho^3}{16}\left(\frac{d}{dz}\right)^3 H(0, z)$$

$$\mp \cdot \cdot + \frac{(-1)^n}{n!\,(n-1)!}\left(\frac{\rho}{2}\right)^{2n-1}\left(\frac{d}{dz}\right)^{2n-1} H(0, z). \tag{11.12}$$

It may be pointed out that the two components H_z and H_ρ are not independent, and must obey div $\mathbf{H} = 0$.

11.5. Find the Scalar Potential of a Current Loop of Radius a at an Observation Point $P(r, \Theta)$, Using Legendre Polynomials

Referring to Fig. 11.3 and remembering from Problem 8 that the potential V along the symmetry axis is known

$$V(0, z) = \frac{I}{2}\left(1 - \frac{z - b}{r}\right) \tag{11.13}$$

or in explicit form,

$$V(0, z) = \frac{I}{2}\left[1 - \frac{z - b}{(c^2 + z^2 - 2cz\cos\alpha)^{1/2}}\right] \tag{11.14}$$

$$\frac{1}{r} = (c^2 + z^2 - 2cz\cos\alpha)^{-1/2}.$$

Inserting $u = \cos\alpha$ in $1/r = (c^2 + z^2 - 2cz\cos\alpha)^{-1/2} = (1/c)[1 + (z^2 - 2czu)/c^2]^{-1/2}$, we expand $1/r$ in a Taylor's series to get

$$\frac{1}{r} = \frac{1}{c}\left[1 + u\left(\frac{z}{c}\right) + \tfrac{1}{2}(3u^2 - 1)\left(\frac{z}{c}\right)^2 + \tfrac{1}{2}(5u^3 - 3u)\left(\frac{z}{c}\right)^3 \right.$$
$$+ \frac{1}{2\cdot 4}(5\cdot 7u^4 - 2\cdot 3\cdot 5u^2 + 1\cdot 3)\left(\frac{z}{c}\right)^4$$
$$+ \frac{1}{2\cdot 4}(7\cdot 9u^5 - 2\cdot 5\cdot 7u^3 + 3\cdot 5u)\left(\frac{z}{c}\right)^5$$
$$+ \frac{1}{2\cdot 4\cdot 6}(7\cdot 9\cdot 11u^6 - 3\cdot 5\cdot 7\cdot 9u^4 + 3\cdot 3\cdot 5\cdot 7u^2$$
$$\left. - 1\cdot 3\cdot 5)\left(\frac{z}{c}\right)^6 + \cdots\right].$$

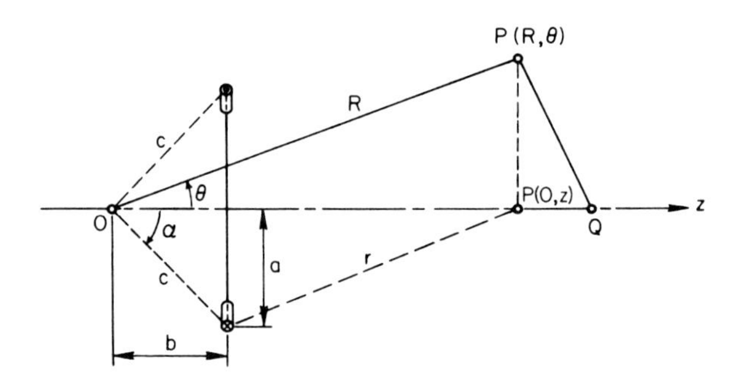

FIG. 11.3.

The coefficient of $(z/c)^n$ is the expression for $P_n(u)$, the Legendre polynomial of the first order. We may write, therefore

$$\frac{1}{r} = \frac{1}{c}\left[1 + \left(\frac{z}{c}\right)P_1(u) + \left(\frac{z}{c}\right)^2 P_2(u) + \ldots + \left(\frac{z}{c}\right)^n P_n(u)\right] \qquad (11.15)$$

with

$$P_n(u) = \sum_{s=0}^{s=m}(-1)^s \frac{(2n-2s)!}{2^n(s!)(n-s)!\,(n-2s)!}\,u^{(n-2s)} \qquad (11.16)$$

where $m = n/2$ or $(n-1)/2$, whichever is an integer.

The scalar potential of the current loop is now written:

$$V(0, z) = \frac{I}{2}\left[1 - \frac{z - b}{c}\sum_{n=0}^{\infty}\left(\frac{c}{z}\right)^{n+1} P_n(\cos\alpha)\right] \qquad z < c. \tag{11.17}$$

Rearranging Eq. (11.17), we get for $z < c$,

$$V(0, z) = \frac{I}{2}\left(1 + \frac{b}{c}\right) - \sum_{n=1}^{\infty}\left(\frac{z}{c}\right)^{n}[P_{n-1}(\cos\alpha) - \cos\alpha P_n(\cos\alpha)]. \tag{11.18}$$

Using the recurrence formula of Legendre polynomials:

$$P_{n-1}(\cos\alpha) - \cos\alpha P_n(\cos\alpha) = \frac{\sin\alpha}{n} P_n'(\cos\alpha) \tag{11.19}$$

$$\cos\alpha P_n(\cos\alpha) - P_{n+1}(\cos\alpha) = \frac{\sin\alpha}{n+1} P_n'(\cos\alpha), \tag{11.20}$$

we get for $z < c$,

$$V(0, z) = \frac{I}{2}\left(1 + \frac{b}{c}\right) - \frac{I\sin\alpha}{2}\sum_{n=1}^{\infty}\frac{1}{n}\left(\frac{z}{c}\right)^{n} P_n'(\cos\alpha). \tag{11.21}$$

To determine the potential $V(R, \Theta)$ at the observation point $P(R, \Theta)$ we replace z_n by R^n and multiply the nth terms of (11.21) by $P_n(\cos\Theta)$ to get:

$$V(R, \Theta) = -\frac{I\sin\Theta}{2}\sum_{n=1}^{\infty}\frac{1}{n}\left(\frac{R}{n}\right)^{n} P_n'(\cos\alpha)\, P_n(\cos\Theta) + \frac{I}{2}\left(1 + \frac{b}{c}\right) \tag{11.22}$$

for $R < c$, remembering that $P_0(\cos\Theta) = 0$.

Example

Calculate the scalar potential of a current loop $V(0, z)$ for all values of $z > c$ and $R > c$.

11.6. Calculate the Field Components of a Current Loop of Radius *a* at *P*(*R*, Θ), Using Legendre Polynomials

We now determine the radial and axial field components B_r and B_Θ of a current loop of radius a at the observation point $P(R, \Theta)$. The flux density components are derived from the scalar potential functions by partial differentiation with respect to R and Θ:

$$B_r(R, \Theta) = \mu_0 \frac{\partial V}{\partial R} = \frac{\mu_0 I \sin \alpha}{2c} \sum_{n=1}^{\infty} \left(\frac{R}{c}\right)^{n-1} P_n'(\cos \alpha)\, P_n(\cos \Theta) \tag{11.23}$$

for $R < c$. Considering that

$$-\frac{\partial}{\partial \Theta} P_n(\cos \Theta) = P_n'(\cos \Theta)$$

we calculate the axial field components

$$B_\Theta(R, \Theta) = -\frac{\mu_0}{r} \frac{\partial V}{\partial \Theta} = -\mu_0 \frac{I \sin \alpha}{2c} \sum_{n=1}^{\infty} \frac{1}{n} \left(\frac{R}{n}\right)^{n-1} P_n'(\cos \alpha)\, P_n'(\cos \Theta) \tag{11.24}$$

for $R < c$.

11.6.1. Remark

If the origin is chosen in the center of the loop, then $\alpha = \pi/2$ and $\cos \alpha = 0$.

Using recurrence formulas for Legendre polynomials, we get for $R < c$:

$$B_r = -\frac{\mu_0 I}{2c} \sum_{n=1}^{\infty} (-1)^{(n+1)/2} \frac{1 \cdot 3 \cdot 5 \cdots n(n+1)}{1 \cdot 2 \cdot 4 \cdot 6 \cdots (n+1)} \left(\frac{R}{c}\right)^{n-1} P_n(\cos \Theta) \tag{11.25}$$

and

$$B_\Theta = \frac{\mu_0 I}{2c} \sum_{n=1}^{\infty} (-1)^{(n+1)/2} \frac{1 \cdot 3 \cdot 5 \cdots n(n+1)}{1 \cdot 2 \cdot 4 \cdots (n+1)n} \left(\frac{R}{c}\right)^{n-1} P_n'(\cos \Theta).$$

$$n = 1, 3, 5, \ldots \quad \text{(odd numbers).} \tag{11.26}$$

11.6.2. Examples

Calculate B_r and B_Θ for $R > c$.

11.6.2.1. Compare B_r and B_Θ obtained in Eqs. (11.25) and (11.26) to Eqs. (11.12) and (11.11).

11.6.2.2. For a current loop of radius $a = 10$ cm, carrying a current of 100 A, calculate B_r and B_Θ using (11.25) and (11.26) at a point P ($r = 3$ cm; $\Theta = 0.5$ rad).

11.6.2.3. Determine lines of constant n ($n = 1, 3, 5, \ldots$) as a function of r and z for the loop given in example 11.6.2.2. Compare the curves with Fig. 13.3.

12. PAIR OF CURRENT LOOPS*

12.1. Problem

Calculate the magnetic field of a pair of circular loops, each having the radius a, and being located at a distance $2l$ from each other. The current flowing through each loop is assumed to be I. The circle planes are parallel (Fig. 12.1). Determine the field distribution inscribed in a sphere of radius r located between the two loops, where $r \leqslant a$.

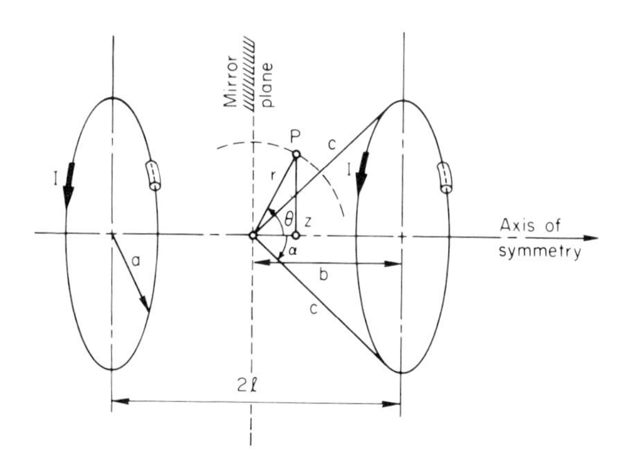

FIG. 12.1.

12.2. Solution

We expect that the magnetic center of the system the axial field curves will have, due to symmetry, a tangent parallel to the symmetry axis. According to the choice of $2l$ and a, we will get either a saddle point or a peak. The main reason for the choice of a pair of loops is to produce an area between the two loops around the symmetry axis, where the field is homogeneous. This means explicitly that over an extended area not only the first derivative of the field components at this area vanishes, but also higher order terms of the field Taylor's expansion disappear.

* Problem 12 is by H. Brechna.

The general configuration of current loops designed to generate such high homogeneous fields is characterized by two types of symmetry:

a. axis of symmetry;
b. mirror plane perpendicular to the axis of symmetry.

We can consider the intersection of the axis of symmetry and the loop pair mirror plane to be the origin of the field coordinates and of the system of loop coordinates.

We are primarily interested in the space around the origin and in the so-called central field region, which is in the interior of a sphere centered at the origin. The ratio l/a must be chosen for a given radius of the sphere, such that the second, fourth, or higher derivatives of field components inside the sphere with prescribed radius r do not exceed certain specified values. Field deviation from the value at the center of the sphere at any point inside the sphere of radius r yields the field uniformity given by the relation $\varepsilon = 1 - [B(r, \Phi, z)/B(0, 0, 0)]$.

By the choice of the location of the mirror plane passing through the origin, all *odd* derivatives of the field at the origin are zero. In order to obtain a homogeneous field, it is thus necessary to eliminate only even derivatives of higher order.

As in previous sections, we again use cylindrical and spherical coordinates, with both sharing a common polar axis (the axis of symmetry) and a common origin (the origin of the system). According to Chapter 11.5, the axial component of the magnetic field along the symmetry axis z for $\Theta = 0$ is given by

$$B_z(z, 0) = \mu_0 I \frac{\sin^2 \alpha}{c} \sum_{m=0}^{\infty} \left(\frac{z}{c}\right)^{2m} P'_{2m+1}(\cos \alpha), \qquad r < c. \tag{12.1}$$

At the observation point $P(r, \Theta)$ the field is expressed as

$$B_z(r, \Theta) = \mu_0 I \frac{\sin^2 \alpha}{c} \sum_{m=0}^{\infty} \left(\frac{r}{c}\right)^{2m} P'_{2m+1}(\cos \alpha)\, P_{2m}(\cos \Theta), \qquad r < c. \tag{12.2}$$

By using the abbreviation

$$B_z(0, 0) = \frac{\mu_0 I \sin^2 \alpha}{c}$$

as the axial field at the origin, the axial field component along the symmetry axis is thus written:

$$B_z(z, 0) = B_z(0, 0)\left[1 + \left(\frac{z}{c}\right)^2 P_3'(\cos\alpha) + \left(\frac{z}{c}\right)^4 P_5'(\cos\alpha) + \cdots\right] \tag{12.3}$$

or in terms of higher order field values:

$$B_z(z, 0) = B_z(0, 0) + B_{2z}(z, 0) + B_{4z}(z, 0) + \cdots + B_{2nz}(z, 0). \tag{12.4}$$

$B_z(0, 0)$ is called the zero order field, $B_2(z, 0)$ the second order field, $B_{4z}(z, 0)$ the fourth order field, and so forth. The terms $B_{2nz}(z, 0)$ are also called successive field corrections. The fields and field corrections explicitly written are:

$$B_z(0, 0) = \mu_0 I \frac{\sin^2\alpha}{c} \tag{12.5}$$

$$B_{2z}(z, 0) = B_z(0, 0)\left(\frac{z}{c}\right)^2 \frac{3}{2}(5\cos^2\alpha - 1) \tag{12.6}$$

$$B_{4z}(z, 0) = B_z(0, 0)\left(\frac{z}{c}\right)^4 \frac{15}{8}(21\cos^4\alpha - 14\cos^2\alpha + 1) \tag{12.7}$$

$$B_{6z}(z, 0) = B_z(0, 0)\left(\frac{z}{c}\right)^6 \frac{3\cdot 7}{2\cdot 4\cdot 6}(3\cdot 11\cdot 13\cdot\cos^6\alpha - 5\cdot 9\cdot 11\cdot\cos^4\alpha + 3\cdot 5\cdot 9\cdot\cos^2\alpha - 1\cdot 5). \tag{12.8}$$

By proper choice of the angle α, or the distance ratio of l/a, we may omit some of the higher order field terms. This elimination yields higher homogeneity. The choice of two current loops will enable us to omit the second order term and we may speak of a second order loop system. Using four loops we may omit the second and fourth order terms, obtaining a sixth order system, etc.

A combination of current loops such that $B_{2z}(z, 0) \equiv 0$ is obtained by choosing α such that $P_3'(\cos\alpha) = \frac{3}{2}(5\cos^2\alpha - 1) = 0$ or $\cos^2\alpha = \frac{1}{5}$, $\alpha = 63° 26'$.

A loop arrangement in this form is also called a Helmholtz combination. A double Helmholtz arrangement will give:

$$B_{4z}(z, 0) \equiv 0$$

or

$$\cos^2\alpha = \frac{1}{3}\left(1 \pm \frac{2}{\sqrt{7}}\right), \qquad \alpha = \pm 49.9°;\ \pm 16.6°.$$

A loop arrangement which satisfies this condition automatically eliminates the second order term.

12.3. Exercise

Express $B_z(z, 0)$ in terms of r and c analogous to Eq. (12.3).

13. FIELD OF A SOLENOID OF FINITE CROSS-SECTIONAL AREA*

13.1. Problem

Calculate the magnetic field of a solenoid of rectangular cross sectional area.

13.2. Solution

The current I flowing through the coil is assumed to be uniform. Design the coil of inner radius a_1 so as to obtain optimum power consumption at maximum field at the coil center. Use coil dimensions given in Fig. 13.1. For detailed analysis we refer to literature.

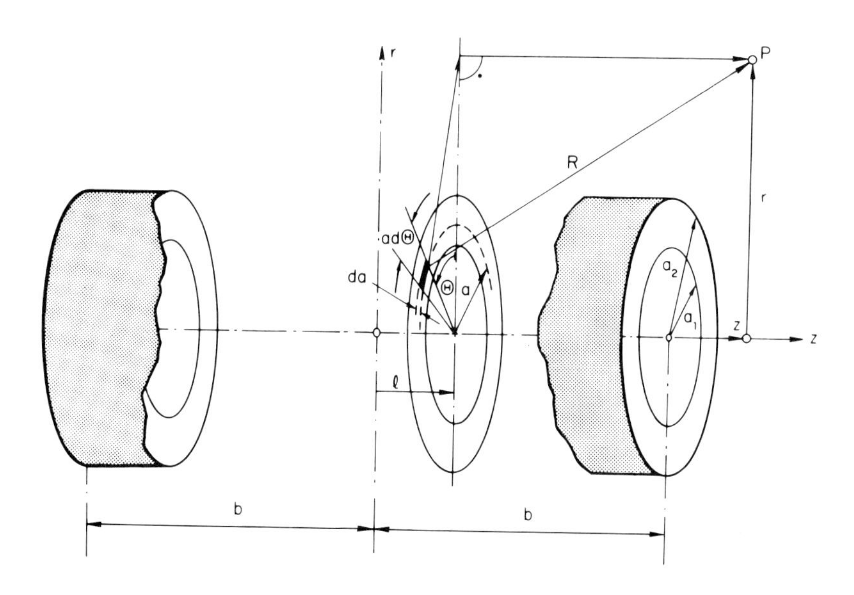

FIG. 13.1.

* Problem 13 is by H. Brechna.

We introduce the definition of current density as the current per unit cross sectional area of the coil. The current flowing through the unit area is thus expressed as:

$$dI = \lambda J \, da \, dz. \tag{13.1}$$

We consider now a current loop with cross sectional area $da\,dz$ and the radius a, and apply Eq. (7.3), assuming the current density over the cross section is constant. The azimuthal component of the magnetic vector potential for a current loop is given by:

$$A_\Phi = \frac{\mu_0}{4\pi} \lambda J \int_{-b}^{+b} dz \int_{a_1}^{a_2} a \, da \int_0^{2\pi} \frac{\cos \Theta \, d\Theta}{[(z-l)^2 + r^2 + a^2 - 2ar \cos_\Theta]^{1/2}}. \tag{13.2}$$

If we introduce the variable $\xi = z - l$ such that at the coil ends, $\xi_1 = z - b$ and $\xi_2 = z + b$, we can simplify the expression for A_Φ to

$$A_\Phi = \frac{\mu_0 J \lambda}{4\pi} \int_{\xi_1}^{\xi_2} d\xi \int_{a_1}^{a_2} a \, da \int_0^{2\pi} \frac{\cos \Theta \, d\Theta}{[\xi^2 + r^2 + a^2 - 2ar \cos \Theta]^{1/2}}. \tag{13.3}$$

The radial and axial field components are obtained by differentiating A_Φ:

$$B_r = -\frac{\mu_0 J \lambda}{2\pi} \int_{a_1}^{a_2} a \, da \int_0^{\pi} \frac{\cos \Theta \, d\Theta}{[\xi^2 + r^2 + a^2 - 2ar \cos \Theta]^{1/2}} \tag{13.4}$$

$$B_z = \frac{\mu_0 J \lambda}{4\pi} \int_{a_1}^{a_2} a \, da \times \int_0^{2\pi} \frac{\xi(a - r\cos\Theta)\, d\Theta}{(a^2 + r^2 - 2ar\cos\Theta)\,(\xi^2 + r^2 + a^2 - 2ar\cos\Theta)^{1/2}} \Bigg|_{\xi_1}^{\xi_2}. \tag{13.5}$$

To solve (13.4) and (13.5) the use of computers is recommended. However, expressions along the symmetry axis may be derived readily from Eqs. (13.4) and (13.5) by setting $r \equiv 0$. Thus $B_r = 0$ and

$$B_z = \frac{\mu_0}{2} J\lambda \{\xi_2 \ln[a + \sqrt{}(\xi_2{}^2 + a^2)^{1/2}] - \xi_1 \ln[a + \sqrt{}(\xi_1{}^2 + a^2)^{1/2}]\} \Big|_{a_1}^{a_2} \tag{13.6}$$

In terms of nondimensional parameters,

$$\alpha = \frac{a_2}{a_1}; \qquad \beta = \frac{b}{a_1}; \qquad \zeta = \frac{z}{a_1}$$

which reduce all coil dimensions to the radius of the bore, we may write for the axial field

$$B_z(0, \zeta) = \frac{\mu_0 J a_1 \lambda}{2} \left\{ (\zeta + \beta) \ln \frac{\alpha + [\alpha^2 + (\zeta + \beta)^2]^{1/2}}{1 + [1 + (\zeta + \beta)^2]^{1/2}} - (\zeta - \beta) \ln \frac{\alpha + [\alpha^2 + (\zeta - \beta)^2]^{1/2}}{1 + [1 + (\zeta - \beta)^2]^{1/2}} \right\}. \tag{13.7}$$

At the center of the coil

$$B_z(0, 0) = \mu_0 J a_1 \beta \lambda \ln \frac{\alpha + (\alpha^2 + \beta^2)^{1/2}}{1 + (1 + \beta^2)^{1/2}}. \tag{13.8}$$

The power consumption of a coil, which produces the field B_z at the center of the coil, is given by:

$$P = 2\pi J^2 \lambda a_1{}^3 \beta (\alpha^2 - 1) \rho \tag{13.9}$$

with ρ the resistivity of the current conducting material and λ the space factor.

If we eliminate J from the two equations, we get the relation between power requirement and central field:

$$B_z(0, 0) = \frac{\mu_0}{(2\pi)^{1/2}} \left(\frac{\beta}{a^2 - 1} \right)^{1/2} \ln \frac{\alpha + (\alpha^2 + \beta^2)^{1/2}}{1 + (1 + \beta^2)^{1/2}} \left(\frac{P\lambda}{a_1 \rho} \right)^{1/2} \tag{13.10}$$

which is commonly written in the form

$$B_z(0, 0) = G(\alpha, \beta) \left(\frac{P\lambda}{a_1 \rho} \right)^{1/2}. \tag{13.11}$$

$G(\alpha, \beta)$ is called the geometry factor.

If B_z is expressed in gauss, a_1 in centimeters, ρ in ohm·cm, power P in watts, then

$$G = \frac{(2\pi)^{1/2}}{5} \left(\frac{\beta}{a^2 - 1} \right)^{1/2} \ln \frac{\alpha + (\alpha^2 + \beta^2)^{1/2}}{1 + (1 + \beta^2)^{1/2}}. \tag{13.12}$$

The geometry factor G as a function of α and β is illustrated in Fig. 13.2.

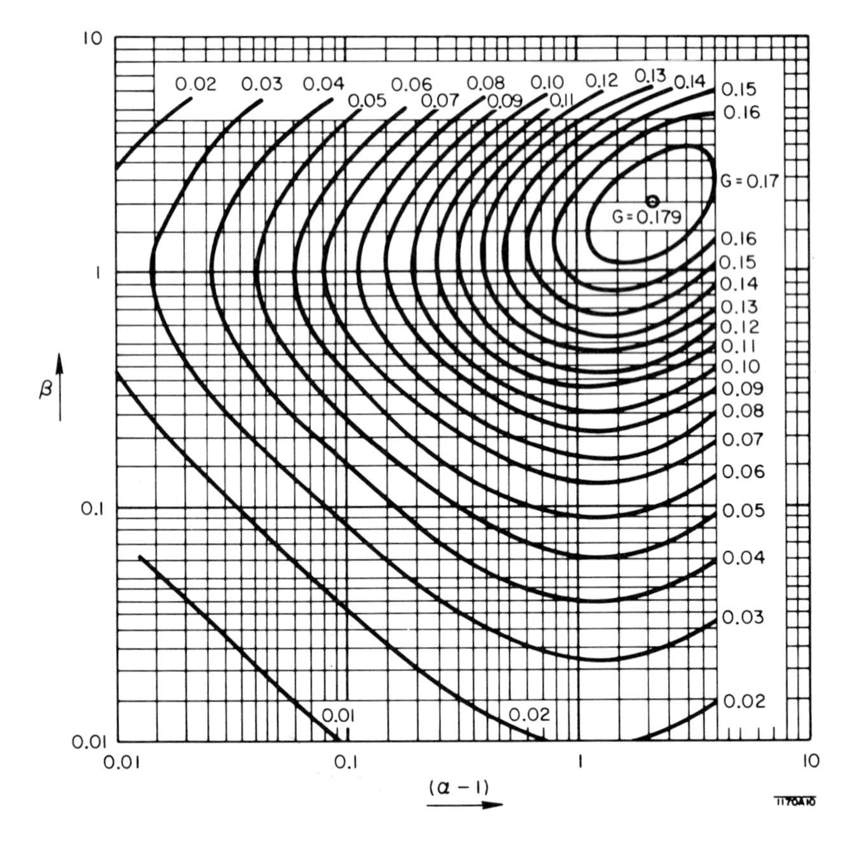

FIG. 13.2.

13.2.1. Example I

Calculate the power of an optimized coil of $a_1 = 5$ cm, producing a field of $B_z = 10$ T $= 10^5$ G at the center. The conductor material is assumed to be copper. Neglect the effect of magneto resistance on copper. In most practical cases the space factor $\lambda = 0.65$, which means that 35% of the cross sectional area is occupied by insulation, structural material, and coolant passages. The resistivity of copper is assumed to be $\rho = 2 \times 10^{-6}$ Ω-cm. From Fig. 13.1 the most optimum value of $G = 0.179$ gives

$$\alpha = 3.2; \qquad \beta = 2.$$

The power requirement of the coil using (13.11) is $P = 48 \times 10^6$ W.

13.2.2. Example II

Calculate the current density in the conductor, the coil ampere turns, and the temperature rise, assuming water cooling or the use of cryogenic liquids.

13.3. Exercise

Calculate the magnetic field of a solenoid of rectangular cross section with uniform current distribution throughout the coil using extensions of higher derivates. Determine the shape or form of the solenoid cross section such that the higher field derivatives vanish.

13.3.1. Solution

In analogy to (11.4.11) and (11.4.12) we can express the field components B_r and B_z in terms of the central field. However, due to symmetry conditions with respect to the origin placed at the center, only even terms appear. Thus:

$$B_z(r, \Theta) = \sum_{n=1}^{\infty} \frac{1}{(2n-2)!} \left[B_z^{(2n-2)}(0, 0) \right] r^{2n-2} P_{2n-2} (\cos \Theta) \tag{13.13}$$

$$B_r(r, \Theta) = \sum_{n=1}^{\infty} \frac{1}{(2n-2)!} \left[B_z^{(2n-2)}(0, 0) \right] r^{2n-2} P'_{2n-2} (\cos \Theta). \tag{13.14}$$

To determine field homogeneity it is important to know the contribution of each term of the series as a function of r and Θ. The influence of the terms is weak in the vicinity of the origin. Each term corresponds to an error function. Envelopes of error coefficients are determined by putting:

$$r^{2n} P_{2n} (\cos \Theta) = \text{const.}$$

and

$$r^{2n} \sin \Theta \, P'_{2n} (\cos \Theta) = \text{const.}$$

In Fig. 13.3 the error contour plots for values of $n = 1$, $n = 2$, and $n = 3$ are illustrated.

The higher order terms of B_z and B_r, as well as the field homogeneity in the vicinity of the origin, are calculated as follows:

Referring to Chapter 13.1 and using the nondimensional parameters α, β, and ζ, we determine the field at any point along the z axis (symmetry) from the expression:

$$B_z = 0.2\pi J a_1 \left\{ (\zeta + \beta) \ln \frac{\alpha + [\alpha^2 + (\zeta + \beta)^2]^{1/2}}{1 + [1 + (\zeta + \beta)^2]^{1/2}} - (\zeta - \beta) \ln \frac{\alpha + [\alpha^2 + (\zeta - \beta)^2]^{1/2}}{1 + [1 + (\zeta - \beta)^2]^{1/2}} \right\}. \quad (13.15)$$

In this equation B_z is given in gauss, if J is expressed in amperes per square

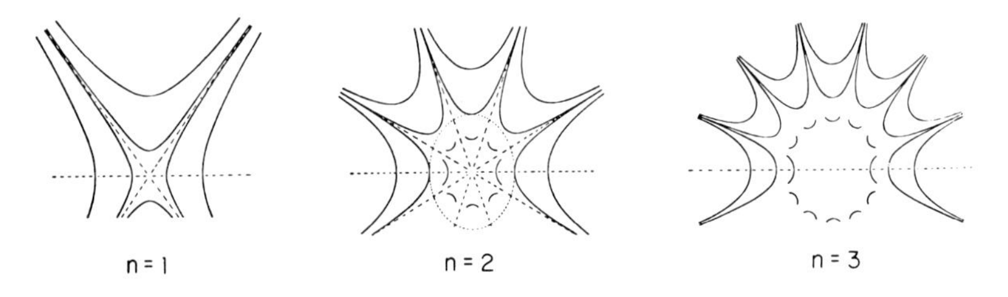

FIG. 13.3.

centimeter and a_1 in centimeters. By introducing the abbreviations

$$C_1 = \frac{1}{1 + \beta^2}; \quad C_2 = \frac{\beta^2}{1 + \beta^2}; \quad C_3 = \frac{\alpha^2}{\alpha^2 + \beta^2}; \quad C_4 = \frac{\beta^2}{\alpha^2 + \beta^2}$$

and differentiating B_z with respect to $\zeta = z/a_1$, show that at the center of the coil, i.e., ($\zeta = 0$),

$$\left(\frac{\partial B_z}{\partial \zeta}\right)_{\zeta=0} = \left(\frac{\partial^3 B_z}{\partial \zeta^3}\right)_{\zeta=0} = \cdots = \left(\frac{\partial^{2n+1} B_z}{\partial \zeta^{2n+1}}\right)_{\zeta=0} = 0. \quad (13.16)$$

The even order derivatives are given by

$$\left(\frac{\partial^2 B_z}{\partial \zeta^2}\right)_{\zeta=0} = \frac{2\pi}{5} J\lambda \frac{a_1}{\beta} (C_1^{3/2} - C_3^{3/2})$$

$$\left(\frac{\partial^4 B_z}{\partial \zeta^4}\right)_{\zeta=0} = \frac{4\pi}{5} J\lambda \frac{a_1}{\beta^3} \left[C_1^{3/2} \left(1 + \frac{3}{2} C_2 + \frac{15}{2} C_2{}^2\right) - C_3^{3/2} \left(1 + \frac{3}{2} C_4 + \frac{15}{2} C_4{}^2\right) \right]$$

$$\left(\frac{\partial^6 B_z}{\partial\zeta^5}\right)_{\zeta=0} = \frac{48\pi}{5} J\lambda \frac{a_1}{\beta^5}\left[C_1^{3/2}\left(1 + \frac{3}{2}C_2 + \frac{15}{8}C_2^{\,2} - \frac{35}{4}C_2^{\,3}\right.\right.$$
$$\left. + \frac{315}{8}C_2^{\,4}\right) - C_3^{3/2}\left(1 + \frac{3}{2}C_4 + \frac{15}{8}C_4^{\,2}\right. \qquad (13.17)$$
$$\left.\left. - \frac{35}{4}C_4^{\,3} + \frac{315}{8}C_4^{\,4}\right)\right]$$

$$\left(\frac{\partial^8 B_z}{\partial\zeta^8}\right)_{\zeta=0} = 288\pi J\lambda \frac{a_1}{\beta^7}\left[C_1^{3/2}\left(1 + \frac{3}{2}C_2 + \frac{15}{8}C_2^{\,2} + \frac{35}{16}C_2^{\,3}\right.\right.$$
$$\left. + \frac{315}{16}C_2^{\,4} - \frac{2079}{16}C_2^{\,5} + \frac{3003}{16}C_2^{\,6}\right)$$
$$- C_3^{3/2}\left(1 + \frac{3}{2}C_4 + \frac{15}{8}C_4^{\,2} + \frac{35}{16}C_4^{\,3} + \frac{315}{16}C_4^{\,4}\right.$$
$$\left.\left. - \frac{2079}{16}C_4^{\,5} + \frac{3003}{16}C_4^{\,6}\right)\right].$$

Using the expression for Taylor's expansion of the field according to (13.13) we have explicitly:

$$B_z(\zeta, \Theta) = \frac{2\pi}{5} J\lambda\, a_2\left[M_0 + M_2\left(\frac{r}{a_2}\right)^2 P_2(\cos\Theta)\right.$$
$$+ M_4\left(\frac{r}{a_2}\right)^4 P_4(\cos\Theta) + \cdots$$
$$\left. + M_{2n}\left(\frac{r}{a_2}\right)^{2n} P_{2n}(\cos\Theta)\right]. \qquad (13.18)$$

with the coefficients

$$M_0 = \beta \ln \frac{\alpha + (\alpha^2 + \beta^2)^{1/2}}{1 + (a \quad + \beta^2)^{1/2}}$$

$$M_2 = \frac{1}{2\beta}(C_1^{\,3/2} - C_3^{\,3/2})$$

$$M_4 = \frac{1}{12\,\beta^3}\left[C_1^{\,3/2}\left(1 + \frac{3}{2}C_2 + \frac{15}{2}C_2^{\,2}\right)\right.$$
$$\left. - C_3^{3/2}\left(1 + \frac{3}{2}C_4 + \frac{15}{2}C_4^{\,2}\right)\right]$$

$$M_6 = \frac{1}{30\,\beta^5}\left[C_1^{\,3/2}\left(1 + \frac{3}{2}C_2 + \frac{15}{2}C_2^{\,2} - \frac{35}{4}C_2^{\,3} + \frac{315}{8}C_2^{\,4}\right)\right.$$
$$\left. - C_3^{3/2}\left(1 + \frac{3}{2}C_4 + \frac{15}{2}C_4^{\,2} - \frac{35}{4}C_4^{\,3} + \frac{315}{8}C_4^{\,4}\right)\right]$$

$$M_8 = \frac{1}{56\,\beta^7}\left[C_1^{\,3/2}\left(1 + \frac{3}{2}C_2 + \frac{15}{8}C_2^{\,2} + \frac{35}{16}C_2^{\,3} + \frac{315}{16}C_2^{\,4}\right.\right.$$
$$\left. - \frac{2079}{16}C_2^{\,5} + \frac{3003}{16}C_2^{\,6}\right) - C_3^{\,3/2}\left(1 + \frac{3}{2}C_4 + \frac{15}{8}C_4^{\,6}\right.$$
$$\left.\left. + \frac{35}{16}C_4^{\,3} + \frac{315}{16}C_4^{\,4} - \frac{2079}{16}C_4^{\,5} + \frac{3003}{16}C_4^{\,6}\right)\right].$$

It is convenient to generate a table of coefficients $M_0 \ldots M_{2n}$ in terms of α and β in order to calculate any higher order terms of the magnetic field.

To eliminate second, fourth, etc., terms of the magnetic field, so-called correcting or compensating coils are added to the main body of the coil as illustrated in Fig. 13.4.

We now express the field of the correcting solenoids in terms of α', β', etc., using the bore diameter of the compensating coil, a_1'. The compensating coil should have the same current density as the main coil, so we may use a single current source. The field of the compensating coil is expressed as above to be:

$$B_z(r, \Theta) = \frac{2\pi}{5} J\lambda\, a_1\left[M_0' + M_2'\left(\frac{r}{a_1}\right)^2 P_2(\cos\Theta)\right.$$
$$+ M_4'\left(\frac{r}{a_1}\right)^4 P_4(\cos\Theta)$$
$$\left. + \cdots + M'_{2n}\left(\frac{r}{a_1}\right)^{2n} P_{2n}(\cos\Theta)\right]. \tag{13.19}$$

To eliminate the second and fourth order terms, we must have:

$$N_2 = M_2(\alpha, \beta) - M_2'(\alpha', \beta') = 0$$
$$N_4 = M_4(\alpha, \beta) - M_4'(\alpha', \beta') = 0.$$

The complicated form of $M'(\alpha', \beta')$ does not permit an exact analytic solution, and thus the method of successive approximation is proposed.

The calculation method, which has proven to be quite useful to obtain sixth order coils, is briefly summarized as

$$B_z(0, 0) = \frac{2\pi}{5} J\lambda\, a_1 N_0(\alpha, \beta, \alpha', \beta') \tag{13.20}$$

which gives the inner coil radius:

$$a_1 = \frac{5}{2\pi} \frac{B_z(0, 0)/J\lambda}{N_0(\alpha, \beta, \alpha', \beta')}. \quad (13.21)$$

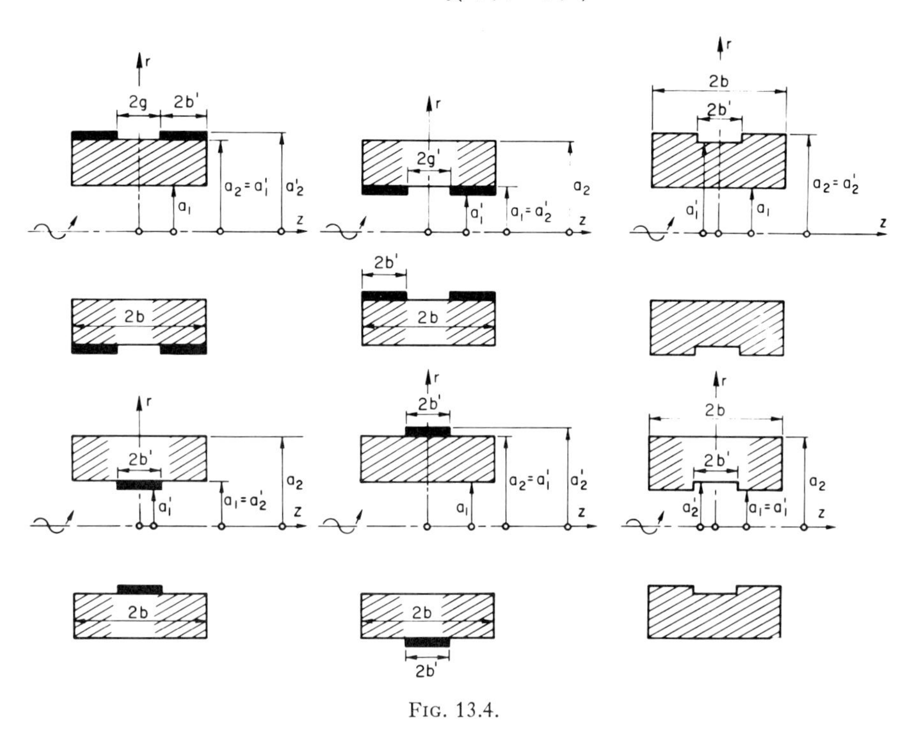

Fig. 13.4.

The volume of the solenoid is given by:

$$V = 2\pi \left(\frac{5B_z(0, 0)/J\lambda}{2\pi N_0}\right)^3 \left[\left(1 - \frac{1}{\alpha^2}\right)\frac{\beta}{\alpha} - \left(1 - \frac{1}{\alpha'^2}\right)\frac{\beta'}{\alpha'}\right]. \quad (13.22)$$

If we introduce the quantity:

$$Z = \frac{125}{4\pi^2 N_0^3}\left[\left(1 - \frac{1}{\alpha^2}\right)\frac{\beta}{\alpha} - \left(1 - \frac{1}{\alpha'^2}\right)\frac{\beta'}{\alpha'}\right], \quad (13.23)$$

the coil volume may be expressed as

$$V = \left(\frac{B_0}{J}\right)^3 Z. \quad (13.24)$$

For a required field and uniform current distribution it is desirable that V, the coil volume, and Z must be a minimum.

The four equations (13.20)–(13.23) must satisfy the conditions

$$
\begin{aligned}
&N_2(\alpha, \beta, \alpha', \beta') = M_2(\alpha, \beta) - M_4'(\alpha', \beta') = 0 \\
&N_4(\alpha, \beta, \alpha', \beta') = M_4(\alpha, \beta) - M_4'(\alpha', \beta') = 0 \\
&Z(\alpha, \beta, \alpha', \beta') \quad \text{must be a minimum.}
\end{aligned}
\tag{13.25}
$$

This system has no analytic solution, as we have three unknowns (α', β' and Z). However, Newton's method may be applied for the first two equations of (13.25) which determine $\alpha'(\alpha, \beta)$ and $\beta'(\alpha, \beta)$. Then we may use approximation methods to find $Z(\alpha, \beta)$.

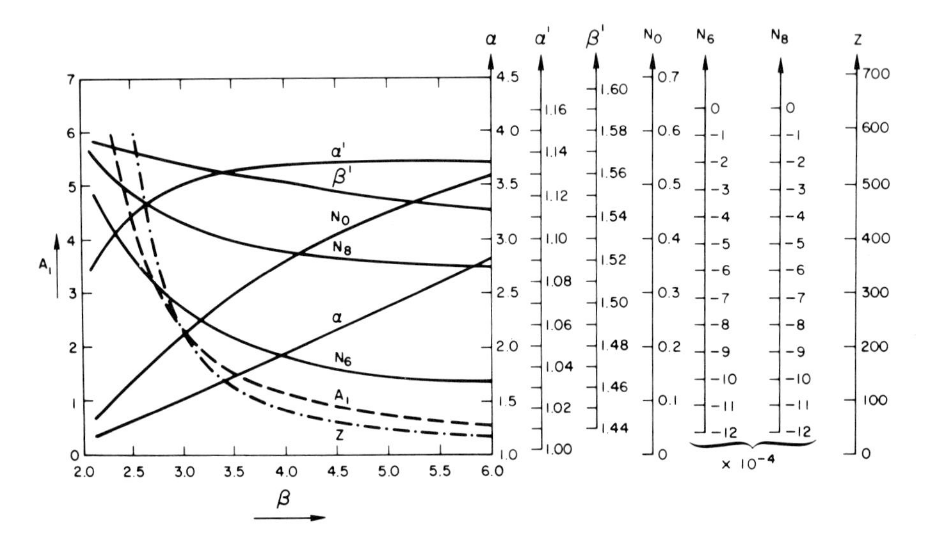

FIG. 13.5.

The nondimensional parameters defining a sixth order coil with minimum volume are given in Fig. 13.5.

The main characteristic parameters of the sixth order coil are:

1. Reduced radial parameter of the principal solenoid $\alpha = a_2/a_1$.
2. Reduced radial parameter of the compensating coil $\alpha' = a_2'/a_1'$.
3. Reduced axial length $\beta' = b'/a_1'$.
4. Constant term of the field expansion N_0.
5. The sixth and eighth order terms N_6 and N_8.
6. The volume coefficient Z.

If $B_0/J\lambda$, the reduced length β, and particularly a_1 are given, the sixth order coil with minimum value can be calculated.

For convenience we have introduced the quantity

$$A = \frac{a_1 J\lambda}{B_0} = \frac{5}{2\pi} \frac{1}{\alpha N_0(\alpha, \beta, \alpha', \beta')}, \tag{13.26}$$

which is plotted in Fig. 13.5.

It may be noted that α', α, β' are functions of β.

13.3.2. Example

A superconducting sixth order coil with a central field of 60 kG, having a space factor $\lambda = 0.6$ and bore radius $a_1 = 3.8$ cm, should be calculated with a minimum coil volume.

Referring to the B–J characteristic of superconductors type II (such as Niobium-Titanium) we find that at a transverse field of 60 kG the current density in the superconductor is 30,000 A/cm^2. The average current density is thus $J \cdot \lambda = 18{,}000$ A/cm^2 with

$$\frac{B_0}{J\lambda} = \frac{60{,}000}{18{,}000} = 3.3333$$

and

$$A = \frac{a_1 J\lambda}{B_0} = \frac{3.8}{3.3333} = 1.1400.$$

This value of A corresponds to a (Fig. 13.5) value of $\beta = 4$. Thus we get

$$\begin{aligned} \alpha &= 1.948 \qquad & N_0 &= 0.3581 \\ \alpha' &= 1.135 & N_6 &= -0.9162 \times 10^{-3} \\ \beta' &= 1.556 & N_8 &= -0.5253 \times 10^{-3} \end{aligned}$$

$$Z = 83.2.$$

The coil dimensions are thus

$$\begin{aligned} a_1 &= 3.8 \text{ cm}; \\ a_2 &= a_1\alpha = 7.4 \text{ cm}; \\ 2b &= 2a_1\beta = 30.4 \text{ cm}; \\ a' &= 6.52 \text{ cm}; \\ 2b' &= 20.29 \text{ cm}. \end{aligned}$$

The axial field is given now by

$$B_z = \frac{2\pi}{5} a_1 J\lambda \left[N_0 + N_6 \left(\frac{z}{a_1}\right)^6 + N_8 \left(\frac{z}{a_1}\right)^8 + \cdots \right]$$

or

$$B_z = B_z(0, 0)\,[1 - 1.55 \times 10^{-8} z^6 + \cdots]$$

with z expressed in centimeters.

The coil volume

$$V = \left(\frac{B_0}{J}\right)^3 Z = 3080 \text{ cm}^3.$$

14. FIELD OF A PAIR OF SOLENOIDS*

14.1. Problem

Determine field distribution and field homogeneity of a pair of solenoids of finite rectangular cross-sectional area. The coil dimensions and the spacing between the solenoids are illustrated in Fig. 14.1.

14.2. Solution

As seen in previous sections, it is an elementary matter to integrate the field produced by a current loop of dimensions $dr\,dz$ with a current density J over the extension of a coil of finite cross sectional area to find the field at any point on the symmetry axis. However, this method does not allow us to find the best possible coil arrangement for the generation of a uniform field subject to reasonable practical constraints. One practical constraint is, for example, that the solenoid cross section be rectangular. For the arrangement illustrated in Fig. 14.1, the winding area is rectangular, with the axial width of $2a_1\beta$ and the radial height $a_1(\alpha' - 1)$.

The coordinates of an element of area $(dr\,dz)$ within the winding space are $(r_1 z)$ or (r, Θ) with respect to the origin at the center of symmetry. (r_0, Θ_0), respectively (r_0, z_0) or $(r_0, a_1\zeta)$ are the coordinates of the center of the winding area.

The field generated by the entire winding is then:

$$B = \mu_0 J \lambda \int_{z_0 - b}^{z_0 + b} dz \int_{a_1}^{a_2} f(r, \Theta, z)\, dr. \tag{14.1}$$

We are interested in fields to be quite uniform over a small volume inside the bore around the origin. We would like to determine what uniformity is obtainable over a small sphere with radius $R \ll a_1$.

For uniformity at the center of the coil we require that as many as possible of the differential field coefficients (see also Chapter 13.3) should vanish inside the sphere of radius a. Since due to symmetry all odd order coefficients are zero, we can ensure that the variations of the field are of fourth orders by requiring that the second differential equation vanishes.

* Problem 14 is by H. Brechna.

As known, four dimensions, g, b, a_1, and a_2, are required to specify the size of the coil. Using these dimensional expressions and the maximum allowable current density J, we can determine the winding space A, the magnetic field B at the center, and the second field derivative noted as D.

We now minimize the area A, subject to the two conditions $B = B_0$, the maximum required field, and $D = 0$, the condition for uniformity. This gives three conditions to be satisfied, and since the gap width $2g$ is

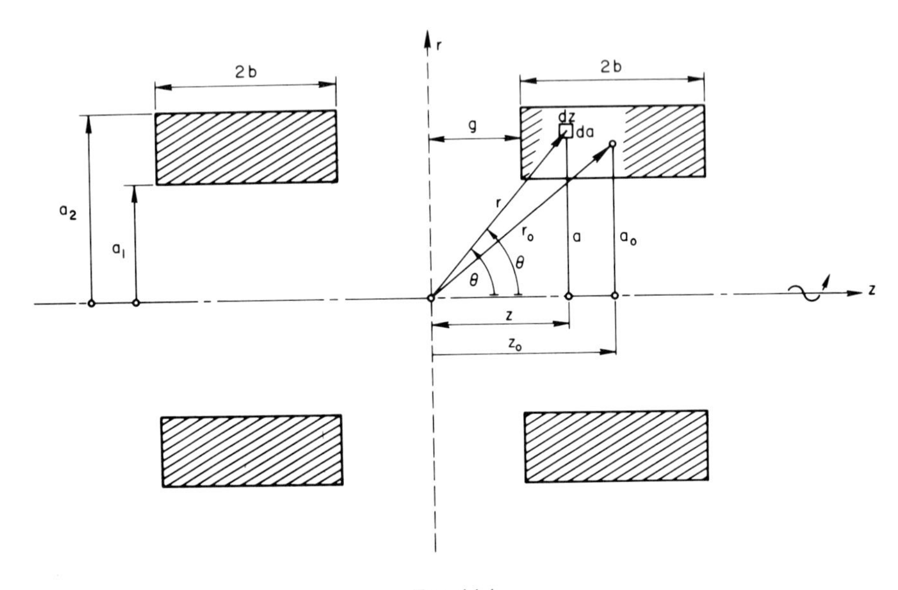

FIG. 14.1.

fixed, due to the experimental requirements (accessibility), we have three parameters to vary, i.e., a_1, a_2, and b.

The total winding space is

$$A = 2\pi(b - g)(a_2{}^2 - a_1{}^2)$$

and the magnetic field at the coil center is given by

$$B = 0.4\pi J\lambda \int_g^b \int_{a_1}^{a_2} \frac{r^2}{(r^2 + z^2)^{3/2}} \, dr \, dz \qquad \text{(gauss).} \tag{14.2}$$

From symmetry, the second differential coefficient at the origin is twice that of one coil. Since differentiation with respect to a field coordinate is equivalent to differentiation with respect to minus the corresponding source coordinate we can obtain D from Eq. (14.2) simply by differentiating with respect to z inside the brackets.

We consider B_0, $J\lambda$, and g as our independent variables instead of a_1, a_2, and b. The dimensions a_1, a_2, and b are now being functions of these new variables.

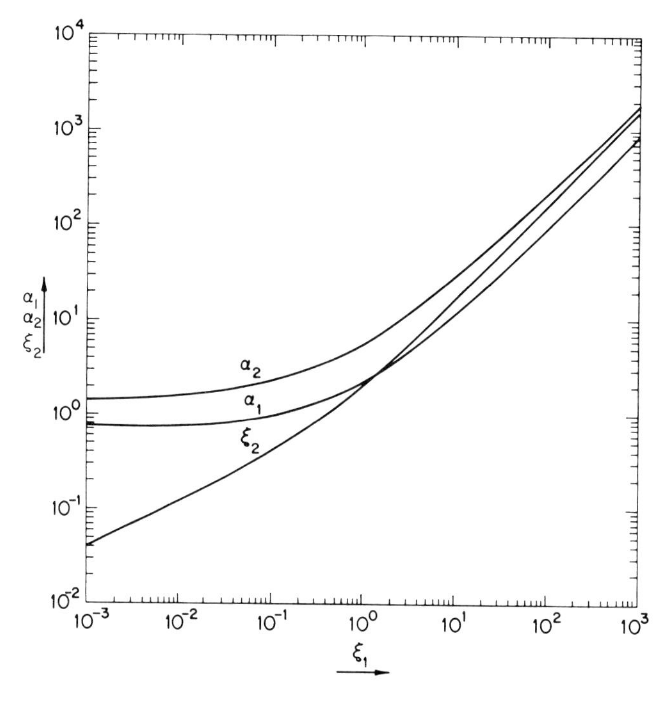

FIG. 14.2.

Now for any given values of B_0, $J\lambda$, and g we can calculate a_1, a_2, and b, as shown below in this section. We consider first B_0 and $J\zeta$. These combine b_0 and $J\lambda$ into one variable, $L = B_0/0.4\pi J\lambda$, and have only two variables, L and g. Since L has the dimension of length, as does the variable g, we make use of dimensionless parameters:

$$\xi_1 = \frac{g}{L}; \qquad \xi_2 = \frac{b}{L}; \qquad \alpha_1 = \frac{a_2}{L}.$$

We see here that ξ_2, α_2, and α_1 are in fact functions of ξ_1 only, i.e., functions of one variable. We can draw graphs of these variables against ξ_1 and calculate the magnet parameter. The terms ξ_2, α_2, and α_1 as functions of ξ_1 are illustrated in Fig. 14.2. To obtain more accuracy we refer to Table I.

TABLE I[a]

$\xi_1 = g \dfrac{\mu_0 J}{B(0,0)}$	$\xi_2 = (2b+g) \dfrac{\mu_0 J}{B(0,0)}$	$\alpha_1 = a_1 \dfrac{\mu_0 J}{B(0,0)}$	$\alpha_2 = a_2 \dfrac{\mu_0 J}{B(0,0)}$
10^{-4}	0.7692	0.01333	1.3335
2×10^{-4}	0.76754	0.01873	1.3470
5×10^{-4}	0.76449	0.02931	1.3737
10^{-3}	0.76249	0.04103	1.4033
2×10^{-3}	0.76177	0.05739	1.4449
5×10^{-3}	0.76592	0.08946	1.5268
10^{-2}	0.77714	0.12559	1.6195
2×10^{-2}	0.80153	0.17766	1.7535
4×10^{-2}	0.84872	0.25483	1.9511
6.4×10^{-2}	0.90164	0.32976	2.1405
8.5×10^{-2}	0.94590	0.38847	2.2870
1.068×10^{-1}	0.98824	0.44294	2.4210
1.885×10^{-1}	1.1383	0.63020	2.8689
3×10^{-1}	1.3247	0.86040	3.3954
4×10^{-1}	1.4812	1.0552	3.8240
1	2.3235	2.1468	6.0383
1.44	2.8190	2.9189	7.4851
2	3.5837	3.8915	9.2231
3	4.7734	5.6200	12.159
4	5.9272	7.3481	14.966
6	8.1741	10.815	20.351
8	10.373	14.301	25.549
10	12.542	17.803	30.625
20	23.162	35.515	55.026
50	54.248	89.754	124.48
10^2	105.33	181.79	236.08
2×10^2	206.69	368.57	453.92
5×10^2	509.06	937.23	1093.2
10^3	1011.4	1895.8	2142.6

[a] Data according to J. D. A. Day, *J. Sci. Instr.* **40,** 583–585 (1963).

It may be remarked that we still need L to give an actual solution; however, because of the homogeneity of the equations this parameter L can be taken as a simple scale factor.

A calculation method is outlined below:

We calculate L and ξ_1 from design data B_0, $J\lambda$, and g. Corresponding values of ξ_2, α_1, and α_2 are obtained from Fig. 14.2. We multiply these values by L and obtain b, α_1, and α_2.

14.2.1. Example

A coil should be designed with the following parameters: $J\lambda = 20{,}000$ A/cm^2; $B_0 = 30{,}000$ G; $g = 3$ cm. From these $L = 1.1937$ and $\xi_1 =$

2.5133. Using Fig. 14.2 we interpolate from values given for $\xi_1 = 1, 2, 3, 4$ to obtain:

$$\xi_2 = 4.2006; \quad \alpha_1 = 4.7792; \quad \alpha_2 = 10.753$$

or

$$b = 5.0141 \text{ cm}; \quad a_1 = 5.7048 \text{ cm}; \quad a_2 = 12.835 \text{ cm}.$$

Of course if more accuracy for a fourth order coil is required, giving a homogeneity of $\sim 10^{-4}$, we may use a suitable computer program and calculate ξ_2, α_1, and α_2 directly.

The same analysis can be applied to the similar problem in which a_1 is specified instead of g. It should be noted that although the condition $(\partial^2 B)/(\partial z^2) = 0$ gives a uniform field at the center, there is no indication of the volume over which the field remains constant. For this we refer to calculations in the next problem.

14.3. Exercise I

Calculate the field homogeneity (sixth order coil) inside a pair of coaxial coils. The radius of the sphere where the homogeneity is required is smaller than the bore radius, or $R \ll a_1$ (Fig. 14.1).

14.3.1. Solution

The field generated by the entire coil (solenoid pair) is written in the form

$$B = \mu_0 J\lambda \int_{-b}^{+b} dz \int_{a_1}^{a_2} f(r, \Theta, z)\, dr. \tag{14.3}$$

We take advantage of the fact that the dimensions $2b$ and $(a_2 - a_1)$ are small compared to the mean coil diameter $2a_0$ and the mean coil spacing $2z_0$.

We can now expand the integral (14.3) in a converging Taylor's series in terms of b and $(a_2 - a_1)/2 = (a_1/2)(\alpha - 1)$.

The only nonvanishing terms of the expansion are those which contain odd powers of $(a_2 - a_1)/2$ or b, or both.

Expanding (14.3) in a Taylor's series yields:

$$B_z = \mu_0 J\lambda 2b(a_2 - a_1)\left\{ f(r_0, \Theta, z) + \frac{b^2}{6}\left(\frac{\partial^2 f}{\partial z^2}\right)_{r_0,\Theta_0} \right.$$

$$+ \frac{(a_2 - a_1)^2}{4 \cdot 6}\left(\frac{\partial^2 f}{\partial a^2}\right)_{r_0,\Theta_0} + \frac{b^4}{120}\left(\frac{\partial^4 f}{\partial z^4}\right)_{r_0,\Theta_0}$$

$$+ \frac{(a_2 - a_1)^2 b^2}{4 \cdot 36} \left(\frac{\partial^4 f}{\partial z^2 \partial a^2} \right)_{r_0, \Theta_0} + \frac{(a_2 - a_1)^4}{16 \cdot 120} \left(\frac{\partial^4 f}{\partial a^4} \right)_{r_0, \Theta_0}$$

$$+ \cdots \Bigg\}. \tag{14.4}$$

In this expansion we may write

$$f(r, \Theta, z) = \left(\frac{\sin^2 \Theta}{r} \right) \sum_{m=0}^{\infty} \left(\frac{z}{r} \right)^{2m} P'_{2m+1}(\cos \Theta) \tag{14.5}$$

for the function f, and its derivatives with respect to cylindrical source coordinates $z = r \cos \Theta$ and $a = r \sin \Theta$ evaluated at the center of the winding r_0, Θ_0.

To find the field generated by a coil of finite cross sectional area, we need only evaluate the derivatives in the form of a series of ascending even powers of the axial field coordinate z. Our aim is now to rewrite the expansion (14.4) of the field as a power series in terms of (z/r_0), namely:

$$B_z = 0.4\pi NI \left(\frac{\sin^2 \Theta_0}{r_0} \right) \left\{ \varepsilon_0 + \varepsilon_2 \left(\frac{z}{r_0} \right)^2 + \varepsilon_4 \left(\frac{z}{r_0} \right)^4 + \cdots \right.$$

$$\left. + \varepsilon_{2n} \left(\frac{z}{r_0} \right)^{2n} \right\} \tag{14.6}$$

where we used the expression for ampere-turns $NI = 2b(a_2 - a_1)J\lambda$ in Eq. (14.5). This is possible only if we can find values of the error coefficients $\varepsilon_2, \varepsilon_4, \ldots, \varepsilon_{2n}$. This requires the evaluation of the derivatives of $f(r, \Theta, Z)$ with respect to the cylindrical source coordinates $z = r \cos \Theta$ and $R = r \sin \Theta$.

The error coefficients appearing in (14.6) are expressed in the form:

$$\varepsilon_0 = 1 + \frac{1}{3} \left(\frac{b}{r_0} \right)^2 P_3' + \frac{1}{4 \cdot 3} \left(\frac{a_2 - a_1}{r_0} \right)^2 \left[\frac{P_3' - P_1'}{5 \sin^2 \Theta_0} - P'_3 \right]$$

$$+ \frac{1}{5} \left(\frac{b}{r_0} \right)^4 P_5' + \frac{1}{16 \cdot 5} \left(\frac{(a_2 - a_1)}{r_0} \right)^4 \left\{ \left[1 - \frac{2}{9 \sin^2 \Theta_0} \right] P_5' \right.$$

$$\left. + \frac{P_3}{\sin^2 \Theta_0} \left[\frac{35}{36} - \frac{2}{3} \right] \right\} + \frac{2}{4 \cdot 3} \left(\frac{(a_2 - a_1) b^2}{r_0^4} \right)$$

$$\times \left(\frac{P_5' - P_3'}{9 \sin^2 \Theta_0} - P_5' \right) + \cdots \tag{14.7}$$

$$\varepsilon_2 = P_3' + 2 \left\{ \left(\frac{b}{r_0} \right)^2 P_5' + \left(\frac{a_2 + a_1}{2r_0} \right)^2 \right.$$

$$\times \left[\frac{1}{9 \sin^2 \Theta_0} (P_5' - P_3') - P_5'\right]\Bigg\}$$

$$+ 3 \left\{\left(\frac{b}{r_0}\right)^4 P_7' + \left(\frac{a_2 - a_1}{2r_0}\right)^4 \left[\left(1 - \frac{2}{13 \sin^2 \Theta_0}\right) P_7'\right.\right.$$

$$\left. + \frac{1}{\sin^2 \Theta_p} \left(\frac{77}{78} - \frac{4}{5}\right) P_5\right] + \frac{10}{3} \left(\frac{(a_2 - a_1)^2 b^2}{4r_0^4}\right)$$

$$\left.\times \left[\frac{1}{13 \sin^2 \Theta_0} (P_7' - P_5') - P_7'\right]\right\} + \cdots \tag{14.8}$$

$$\varepsilon_4 = P_5' + 5 \left\{\left(\frac{b}{r_0}\right)^2 P_7' + \left(\frac{a_2 - a_1}{2r_0}\right)^2\right.$$

$$\left.\times \left[\frac{1}{13 \sin^2 \Theta_0} (P_7' - P_5') - P_7'\right]\right\} + \cdots \tag{14.9}$$

$$\varepsilon_6 = P_7' (\cos \Theta_0) + \cdots \quad . \tag{14.10}$$

We now recall that the uniformity of the magnetic field in the central field region can be made dependent on the cancellation of the successive even powers of z in the expansion of the field at the origin. For one single pair of solenoids (Helmholtz arrangement) the angle Θ_0 must be chosen such that $P_3' (\cos \Theta_0) = 0$. Therefore

$$\cos \Theta_0 = \frac{1}{\sqrt{5}} \cdot \tag{14.11}$$

As we have seen in Problem 12, this condition will cause the second power to vanish for an infinitesimal cross section.

For a finite rectangular cross sectional area, however, the second error coefficient is

$$\varepsilon_2 = 2P_5' (\cos \Theta_0) \left[\left(\frac{b}{r_0}\right)^2 - \left(1 - \frac{1}{9 \sin^2 \Theta_0}\right) \left(\frac{a_2 - a_1}{2r_0}\right)^2\right.$$

$$\left. - \frac{1}{2 \sin^2 \Theta_0} \left(\frac{77}{78} - \frac{4}{5}\right)\right] + P_7' \left[3 \left(\frac{b}{r_0}\right)^4\right.$$

$$\left. + \left(1 - \frac{2}{13 \sin^2 \Theta_0}\right) \left(\frac{a_2 - a_1}{4r_0}\right)^4 + \cdots\right].$$

In order to make ε_2 approach zero, the part

$$\left(\frac{b}{r_0}\right)^2 - \left(1 - \frac{1}{9 \sin^2 \Theta_0}\right) \left(\frac{a_2 - a_1}{2r_0}\right)^2 - \frac{1}{2 \sin^2 \Theta_0} \left(\frac{77}{78} - \frac{4}{5}\right) = 0.$$

Neglecting the term independent of the coil cross sectional dimensions, we get for

$$P_3' \cos \Theta_0 = 0; \qquad \cos \Theta_0 = \frac{1}{\sqrt{5}}; \qquad \sin \Theta_0 = \frac{2}{\sqrt{5}}$$

that

$$\frac{2b}{a_2 - a_1} = \left(1 - \frac{1}{9 \sin^2 \Theta_0}\right)^{1/2} = 0.928. \tag{14.12}$$

The next terms in the expansion of ε_2 are of the fourth order in $(a_2 - a_1)/r_0$ and b/r_0. The error term is ε_4; the filamentary current loop pair is modified for the second order terms in b/r_0 and $(a_2 - a_1)/r_0$ by $P_5' = (\cos \Theta_0) = -\frac{9}{5}$.

The finite cross sectional area modifies the simple fourth order Helmholtz field by terms of the sixth order in b/r_0, $(a_2 - a_1)/r_0$, and z/r_0 combined, when the condition (14.12) is fulfilled.

14.3.2. Example I

Express B_z in terms of $(z/a_1)^{2n}\ \varepsilon'_{2n}$ as shown in (14.6). Find the relation between ε'_{2n} and ε_{2n}.

14.3.3. Example II

Improperly spaced solenoids. Given a coil configuration as follows: $a_1 = 15$ cm, $a_2 = 45$ cm, $g = 10$ cm, $2b = 29$ cm, $NI = 2.775 \times 10^6$ for one solenoid.

The constant part of the axial field at the origin is calculated from the expression:

$$B_0 = \mu_0 NI\ \varepsilon_0 \frac{\sin^2 \Theta_0}{r_0}.$$

In this example we find that $B_0 = 52.150$ kG.

The second and fourth error coefficients calculated from Eqs. (14.7)–(14.10) are accordingly:

$$\varepsilon_2' = 1.99 \times 10^{-1}, \quad \varepsilon_4' = -5.70 \times 10^{-2}, \quad \varepsilon_6' = -4.02 \times 10^{-3}.$$

The axial field of the coil upon the origin is given by

$$B_z(0,0) = 52.15 \times 10^3 \left[1 + 1.99 \times 10^{-1} \left(\frac{z}{a_1}\right)^2 - 5.7 \times 10^{-2} \left(\frac{z}{a_1}\right)^4 - 4.02 \times 10^{-3} \left(\frac{z}{a_1}\right)^6 + \cdots\right] \quad \text{(gauss)}.$$

Generally, the area where the field should be homogeneous within certain limits is much smaller than a_1. For one particular experiment it was requested that $z = 1.27$ cm. In this case:

$$B_z = 52.15 \times 10^3 \,[1 + 1.42 \times 10^{-3} - 2.92 \times 10^{-6} - 2.92 \times 10^{-10} + \cdots].$$

The second order term is dominant and equal to 74 G, and for many applications not suitable. Two ways are possible to omit the second order term:

(1) $P_3'(\cos \Theta_0) \equiv 0$, therefore $\cos \Theta_0 = 1/\sqrt{5}$ and $\tan \Theta_0 = 2$; $\tan \Theta_0 = (a_1 + a_2)/2(b + g) = 30/(14.5 + g) = 2$, $g = 0.5$ cm (too small).

(2) $b/a = 0.928$, from Eq. (17.22), i.e., $b = 0.928 \cdot a_1(\alpha - 1)/2$.

We assume a_1 and α are given: $a_1 = 15$ cm; $\alpha = 3$, $b = 13.92$ cm instead of 14.5 cm.

14.3.4. Remark

For detailed study we refer to literature, specifically to the work of Garrett and Gauster, and to tables worked out by Montgomery and Terrell.

14.4. Exercise 2

Design a volume minimized-split solenoid system of high field homogeneity. Use Fig. (14.2) for reference.

14.4.1. Solution

In this chapter we modify the field equation (13.7) such that either graphs or calculation by means of computers may be utilized.

Equation (13.7) can be modified as:

$$F(0) = \beta_1 \ln \frac{\alpha + [\alpha^2 + \beta_1{}^2]^{1/2}}{1 + [1 + \beta_1{}^2]^{1/2}} - \beta_2 \ln \frac{\beta + [\alpha^2 + \beta_2\ ]^{1/2}}{1 + [1 + \beta_2{}^2]^{1/2}}. \tag{14.13}$$

In this equation,

$$\beta_1 = \frac{2b + g}{a_1}; \qquad \beta_2 = \frac{g}{a_1}.$$

The volume of the split coil system is written as

$$V = a_1{}^3 v \tag{14.14}$$

with

$$v = 2\pi(\beta_1 - \beta_2)(\alpha^2 - 1) \tag{14.15}$$

as the reduced coil volume.

The magnetic field at the origin is expressed as:

$$\begin{aligned} B_z(0, 0) &= \mu_0 J \lambda a_1 F(0) \\ &= \mu_0 J \lambda a_1 [F_1(0) - F_2(0)]. \end{aligned} \tag{14.16}$$

The constraint of volume minimization follows from

$$\frac{\partial F(0)}{\partial \alpha} = 0 \tag{14.17}$$

for v, β_2 fixed.

The final free variable, β_2, is fixed by specifying the axial field homogeneity over some portion of the gap. There we may use zonal harmonic expansion techniques developed by Garrett. The axial field can be expanded in reduced form about the origin on axis and in the plane of symmetry as follows:

$$\kappa = \sum_{n=0}^{\infty} \varepsilon_{2n} \left(\frac{r}{a_1}\right)^{2n} P_{2n} (\cos \Theta). \tag{14.18}$$

ε_{2n} is as usual the $2n$th error coefficient for the system, written as

$$\varepsilon_{j\,2n} = \frac{K_1(0)\ \varepsilon_{1\,2n} - K_2(0)\ \varepsilon_{2\,2n}}{K_1(0) - K_2(0)}, \qquad j = 1, 2, 3 \ldots \tag{14.19}$$

ε_{12} = the second order error coefficient for coil of length $4b + 2g$

ε_{22} = the second order error coefficient for coil of length $2g$.

By using the expression $\zeta = z/a_1$; $\xi = a/a_1$, we can rewrite the geometry functions:

$$\begin{aligned} F_j(\alpha; \beta_j; \zeta) = \frac{1}{2} \Bigg\{ &(\beta_j + \zeta) \ln \frac{\alpha + [\alpha^2 + (\beta_j + \zeta)^2]^{1/2}}{1 + [1 + (\beta_j + \zeta)^2]^{1/2}} \\ &+ (\beta_j - \zeta) \ln \frac{\alpha + [\alpha^2 + (\beta_j - \zeta)^2]^{1/2}}{1 + [1 + (\beta_j - \zeta)^2]^{1/2}} \Bigg\}. \end{aligned} \tag{14.20}$$

The second order error coefficient is written:

$$\varepsilon'_{j\,2n} = \frac{1}{(2n)!F_j(0)} \quad \frac{\partial^{(2n)} F_j(\alpha\ \beta_j, \zeta)}{\partial \zeta^{2n}}. \tag{14.21}$$

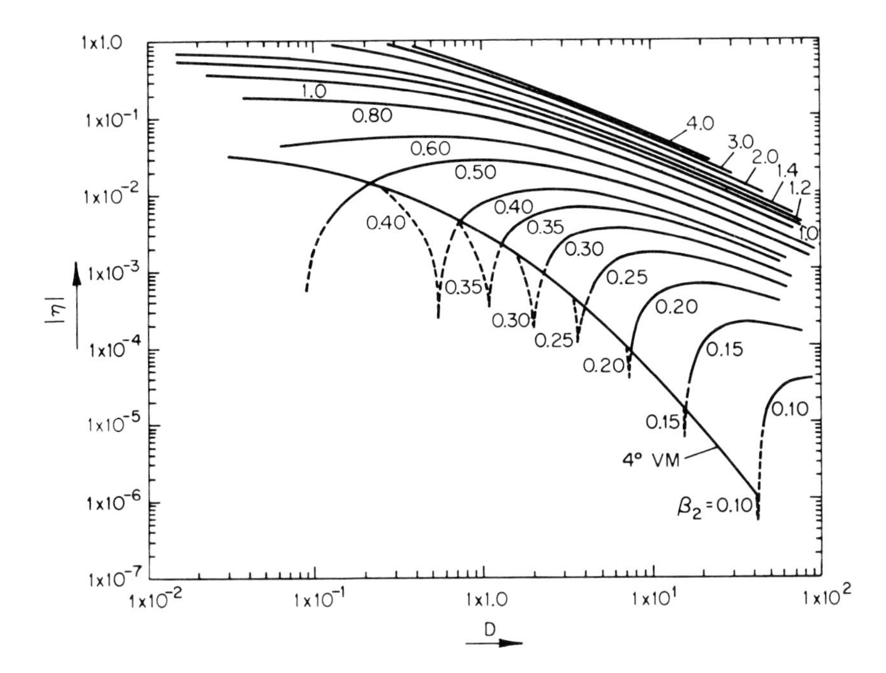

Fig. 14.3.

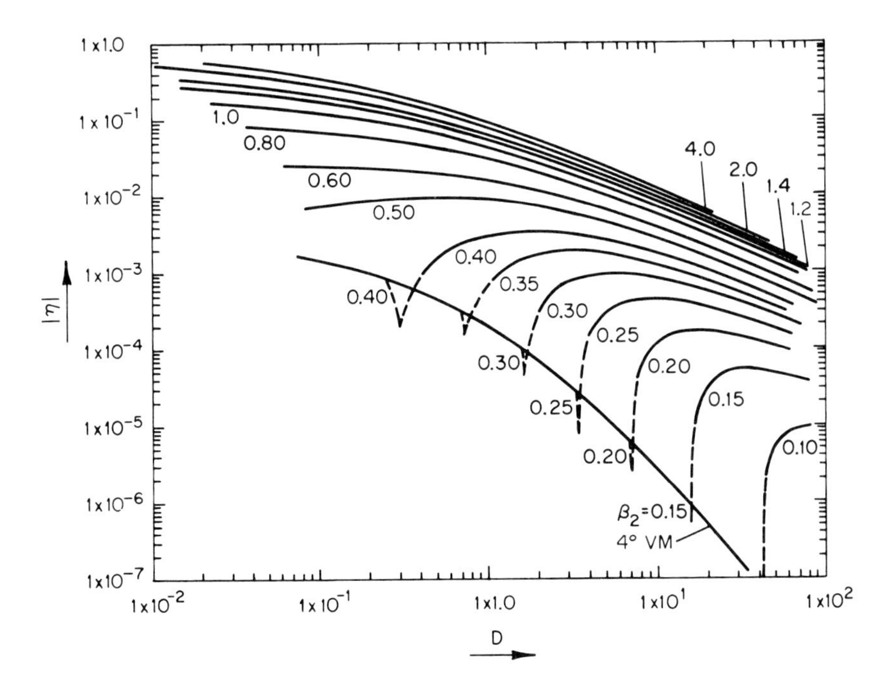

Fig. 14.4.

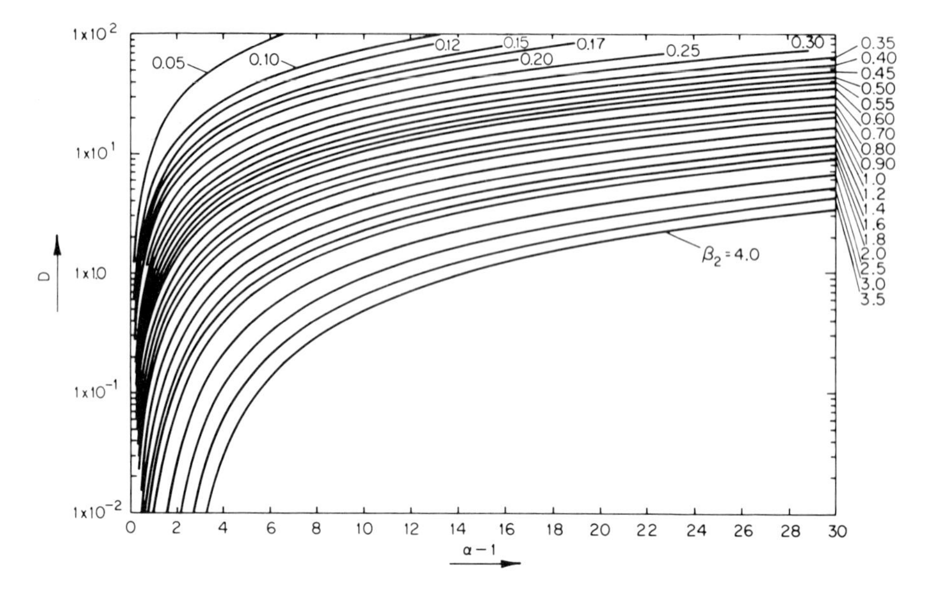

Fig. 14.5.

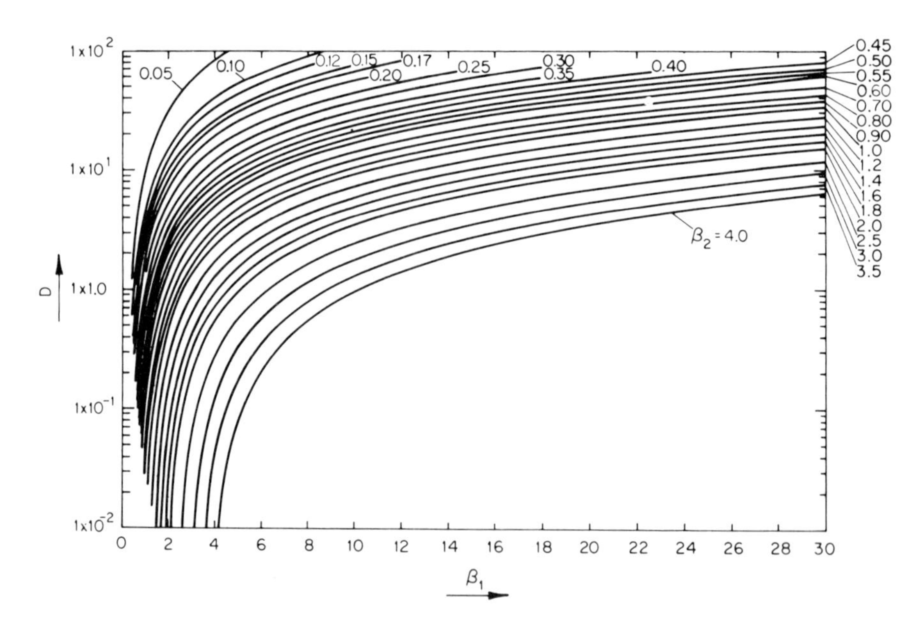

Fig. 14.6.

Having obtained ε_{2n}, we can write the homogeneity of the axial field along the z axis as:

$$\begin{aligned}\eta_\zeta &= \kappa - 1 \\ &= \sum_{n=1}^{\infty} \varepsilon_{2n}\, \zeta^{2n}; \end{aligned} \tag{14.22}$$

and in the plane of symmetry:

$$\eta_\xi = \sum_{n=1}^{\infty} (-1)^n \frac{(2n-1)!}{(2n)!} \varepsilon_{2n} \xi^{2n}.$$

14.4.2. Example I

Fourth order design for fixed β_2. Find α and β_1 such that:

$$\frac{\partial F}{\partial \alpha} = 0 \qquad \text{for} \quad (v_{\text{min}};\ F_{\text{max}}), \qquad \varepsilon_{22} = 0.$$

This is a Helmholtz solution.

The condition expressed in (14.17) for β_2 and v fixed results in a unique point in the (β_1, α) plane. If v is varied a line is generated, where $(\partial F/\partial \alpha) = 0$ holds. Similarly, the vanishing of (14.19) results in another line in the same plane (α, β_1). The intersection of both lines yields the desired solution.

14.4.3. Example II

Given β_2 and v, find α and β_1 such that

$$\frac{\partial F}{\partial \alpha} = 0 \qquad (v_{\text{min}};\ F_{\text{max}}).$$

This is a constraint of volume minimized coil in Example 1. We compute η. We define:

$$D = \frac{B_0}{\mu_0 J \lambda g} = \frac{F(0)}{\beta_2} \tag{14.23}$$

and use D as the basic variable, obtained from specified quantities. The homogeneity η versus D is illustrated for $\zeta = \beta_2$ in Fig. 14.3 and for $\zeta = 0.5\beta_2$ in Fig. 14.4.

To give a numerical example we specify the following magnet: $B(0, 0) = 100$ kG; $\lambda J = 5 \times 10^3$ A/cm^2; $g = 4$ cm. From (17.23) we get $D = 3.978$. For a fourth order coil Fig. 14.4 gives $\eta = 3 \times 10^{-4}$ $(\zeta = \beta_2)$ and from Fig. 14.4: $\eta = 2 \times 10^{-5}$ $(\zeta = 0.5\beta_2)$. From Fig. 14.3 we also obtain the value of $\beta_2 = 0.25$. We calculate $a_1 = g/\beta_2 = 16$ cm. For constant values of β_2 we have computed D versus $\alpha - 1$ in Fig. 14.5, which gives us for the above magnet $\alpha = 2.9$. The corresponding value of β_1 is obtained from Fig. 14.6, $\beta_1 = 1.88$, which gives us the reduced volume of $v = 75.89$ from Eq. (14.15).

SUGGESTED REFERENCES

1. F. Bitter, The design of powerful electromagnets, Part I; The magnetizing coil, Part II, *Rev. Sci. Instr.* **7,** 479–488 (1963).
2. J. P. Blewett, Magnetic field configurations due to air-core coils, *J. Appl. Phys,* **18,** 968–982 (1947).
3. H. Brechna, and D. B. Montgomery, A High Performance dc Magnet Utilizing Axial Cooled Discs, Nat. Magnet Lab., NML-62-1 (1962).
4. H. Brechna, *Electromagnets. Sect. III.* In preparation.
5. J. D. A. Day, Superconducting magnet design, *J. Sci. Instr.* **40,** 583–585 (1963).
6. W. Franzen, Generation of uniform magnetic fields by means of air-core coils, *Rev. Sci. Instr.* **33,** 933–938 (1962).
7. M. W. Garrett, Axially symmetric systems for generating and measuring magnetic fields, Part I, *J. Appl. Phys.* **22,** 1091–1107 (1951).
8. F. Gaume, New solenoid magnets, *Proc. Intern. Conf. High Magnetic Fields,* p. 27–38. MIT Press, Cambridge, Massachusetts; and Wiley, New York, 1961.
9. W. F. Gauster, Some basic concepts for magnet coil design, *Trans. Am. Inst. Elec. Engr.*, Part I; *Commun. Electron,* **79,** 822–828 (1961).
10. B. Girard, and M. Sausade, Calcul des solenaides compenses du 6ème ordre A volume de bobinage minimum, *Nucl. Instr. and Methods,* **25,** 269–284 (1964).
11. H. C. Hitchcock, The design of volume minimized split solenoid systems, *Proc. Intern. Symp.* Magnet Tech., Conf. 650922, pp. 111–125 (1965).
12. D. B. Montgomery and J. Terell, Some useful information for the design of air-core solenoids, Nat. Magnet Lab. AFOSR–1525 (1962).
13. W. R. Smithe, "*Static and Dynamic Electricity,*" pp. 134–140. McGraw-Hill, New York, 1950.

15. BROAD BAND IMPEDANCE MATCHING*

15.1. Problem

A device is to be supplied power from a 50-Ω source over the frequency range 400–800 MHz. The input to the device is by a standard coaxial type-N female connector (shield I. D. = 8.4 mm). It is desired that all of the power from the source (of 50-Ω source impedance) is absorbed in the device. However, since the impedance of the device is not constant over the 400–800 MHz range, this is recognized to be impossible. Instead of complete power absorption, an arbitrary limit of 90% absorption, 10% reflection is considered tolerable for the particular case in question.

It is obvious that without a knowledge of the impedance behavior of the device, it will be impossible to design any system for meeting the requirement. Consequently, the following results (Table I) of standing wave measurements have been obtained for the device

TABLE I. Standing Wave Measurements

Frequency (MHz)	Position of voltage minimum in front of plane A–A (see Fig. 15.1) (cm)	VSWR
400	9.30	2.45
500	9.60	1.60
600	9.60	2.40
700	7.82	3.50
800	5.70	3.40

Design a system that meets the 90–10% requirement stated above.

15.2. Solution

In order to satisfy the stated requirement, we must design a broad band impedance matching network. The characteristics of this network are

* Problem 15 is by I. Kaufman.

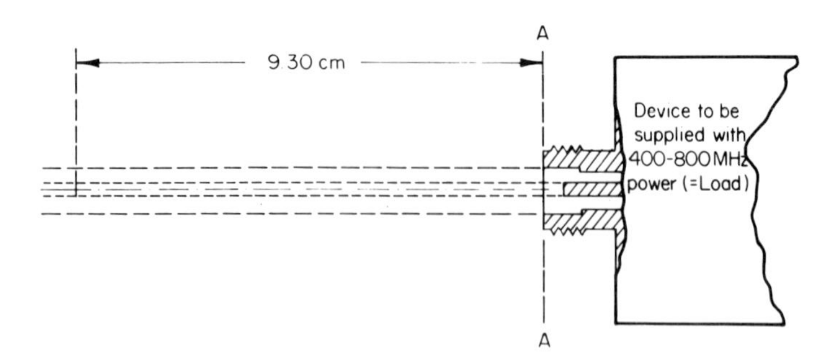

FIG. 15.1. When measured with a slotted line at 400 MHz, the first voltage minimum lies 9.30 cm in front of plane A-A (air dielectric). Impedance of slotted line is 50 Ω.

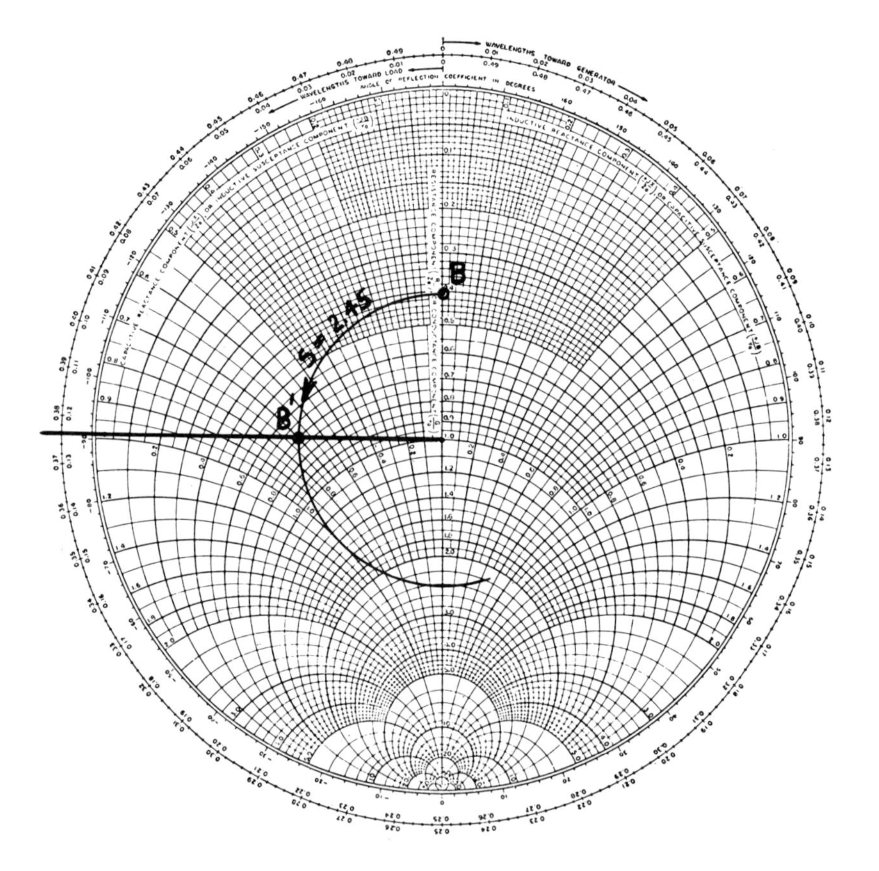

FIG. 15.2. Method of determining impedance at plane A-A from S and position of voltage minimum.

derived from the impedance behavior of the device. This can be found from the VSWR (voltage standing wave ratio) and the position of the voltage minimum. While the impedance at each frequency could be calculated by use of the transmission line equations, it is much easier to use a Smith chart. An example is given for the 400-MHz point.

At 400 MHz, the free space wavelength λ_0 is 75.0 cm. Therefore 9.30 cm $= 0.124\ \lambda_0$. Letting S be the VSWR, and Z_0 be the characteristic impedance (50 Ω), the impedance at the voltage minimum, looking toward the load, is $Z_{\min} = Z_0/S$. For 400 MHz, therefore, $Z_{\min} = Z_0/2.45 = 0.409\ Z_0 + j0$.

We plot this point on the Smith chart (point B) Fig. 15.2. The impedance at Plane A-A is found by following the circle that corresponds to $S = 2.45$ "toward the load", an angle corresponding to $0.124\ \lambda_0$. The result is point B', whose normalized impedance, read off the Smith chart, is $0.70 - j0.70$. The actual impedance is, of course, obtained by multiplying this number by 50 Ω.

The same procedure is used for the other frequencies. The result follows in Table II.

TABLE II.

Frequency (MHz)	λ_0 (cm)	Distance of voltage minimum to load, in wavelengths of the frequency concerned (λ_0)	Load impedance, normalized to 50 Ω	Actual load impedance (Ω)
400	75	0.124	$.70 - j\,0.70$	$35 - j\,35$
500	60	0.160	$1.10 - j\,0.49$	$55 - j\,24.5$
600	50	0.192	$1.50 - j\,0.98$	$75 - j\,49$
700	42.8	0.183	$1.20 - j\,1.44$	$60 - j\,72$
800	37.5	0.152	$0.75 - j\,1.10$	$37.5 - j\,55$

To follow the procedure of impedance matching, let us now resort to a rectangular plot. This is Fig. 15.3. Here the impedance of the load is the lowest curve on the figure. It was plotted from the data just arrived at by use of the Smith chart.

The specifications state that at least 90% of the incident power from the source is to be absorbed in the device. Since the ratio of reflected to incident power is $|\Gamma|^2$, where Γ is the load reflection coefficient, we have here the acceptable limit of $|\Gamma_{\lim}|^2 = 0.1$; so that $|\Gamma_{\lim}| = 0.316$.

Consequently, the acceptable maximum standing wave ratio is

$$S_{\text{lim}} = \frac{1 + |\Gamma_{\text{lim}}|}{1 - |\Gamma_{\text{lim}}|} = 1.92.$$

The load impedance is therefore allowed to fluctuate in the circle whose intercepts with the R axis of Fig. 15.3 are 50/1.92 and (50) (1.92), i.e. 26.0 and 96.1 Ω, respectively. This circle, which is generally termed the "definition circle", is shown in the center of Fig. 15.3. We note that most of the impedance locus of our device lies outside of this definition circle. To meet the specifications, we must therefore add some form of compen-

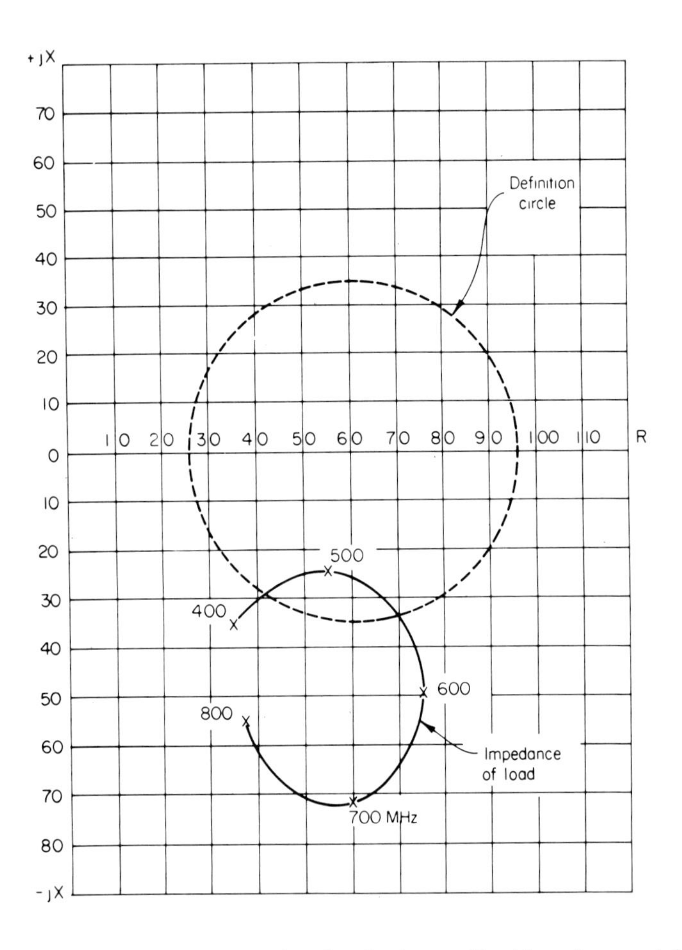

FIG. 15.3. Rectangular impedance plot, showing locus of load impedance and definition circle.

sating network that shifts the compensated impedance entirely into the definition circle.

The particular impedances corresponding to the load impedance locus here are found to have resistance components that lie entirely within the span of the 26.0–96.1 Ω of the definition circle. It is only the reactances that are incorrect. Consequently, we need to cancel some of the large negative reactive components with positive reactances. The simplest arrangement is a network of series reactance.

Since the frequencies here are too high for compensation by lumped elements, we resort to transmission line techniques, using the circuit arrangement of Fig. 15.4.

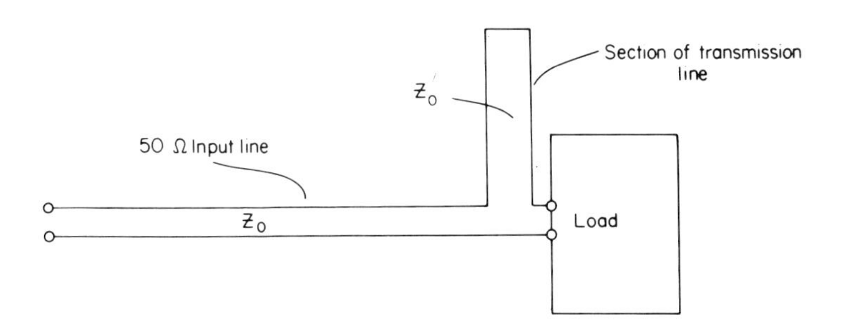

FIG. 15.4. Compensation by a section of transmission line in series with the load.

We consider a shorted transmission line as the compensating element. Suppose the length of this line is chosen to be less than $\lambda_0/4$ at 600 MHz and of such an impedance level that its impedance at 600 MHz is $+j49$. It thus cancels the reactive component of the load at 600 MHz completely. Now, the value of this series reactance is $X = Z_0' \tan \theta$; where Z_0' is the characteristic impedance of the compensating line; θ is the length of the compensating line. (If the length of this line is L, then $\theta = 2\pi L/\lambda_0$.) At frequencies above 600 MHz, θ exceeds the 600-MHz value. If it remains below $\pi/2$ rad, X will be increased. The converse is true at lower frequencies. A glance at Fig. 15.3 will reveal that this is just what is needed.

Consequently, we try compensation with a shorted section of transmission line, placed in series, as in Fig. 15.4.

In our first attempt, we will let $Z_0' = 50\ \Omega$. We choose the length of the series line such that at 600 MHz, the compensated impedance is $75 + j0$. At 600 MHz, therefore, $Z_0' \tan \theta = +j49$. Thus $\tan \theta = 49/50 = 0.98$, and $\theta = 44.4°$. Results for the other frequencies are given in Table III.

TABLE III.

Frequency (MHz)	θ (degrees)	$\tan\theta$	$Z_0' \tan\theta$	Compensated impedance
600	44.4	0.98	49	$75+j0$
400	29.6	0.57	28.5	$35-j6.5$
500	37.0	0.75	37.5	$55+j13$
700	51.8	1.26	63	$60-j9$
800	59.2	1.68	84	$37.5+j29$

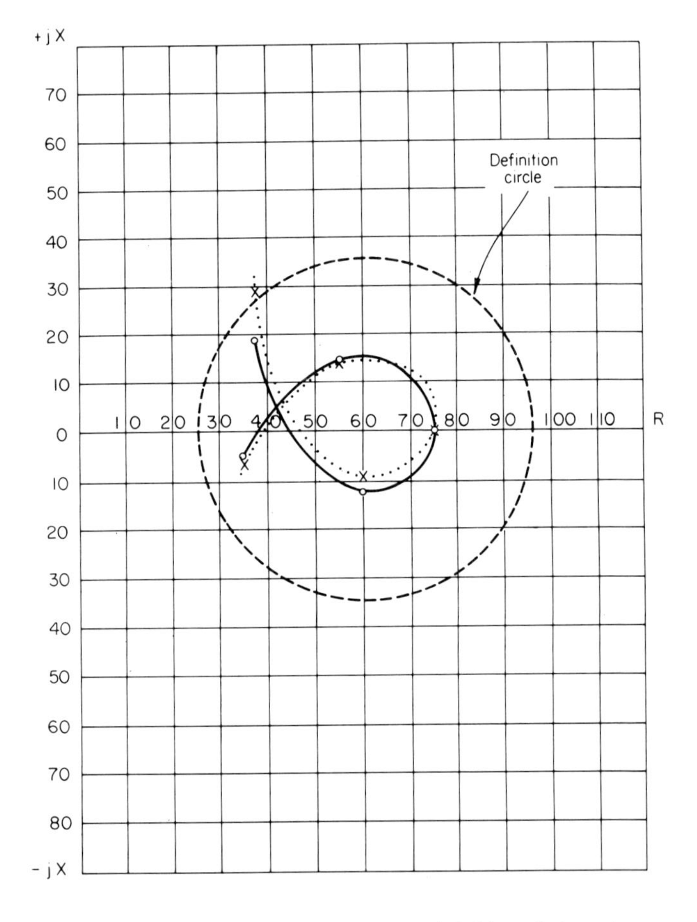

FIG. 15.5. Rectangular impedance plot, showing definition circle and curves of compensated impedance. Dotted curve: initial attempt; solid curve: second attempt.

The values of compensated impedance are plotted on the rectangular impedance diagram, Fig. 15.5. We note that the locus of this compensated impedance does not quite lie within the definition circle.

In our second attempt, we try a transmission line of somewhat higher Z_0', say, $Z_0' = 70\ \Omega$. Again, we let $Z_0' \tan \theta = 49$ at 600 M_2. Then $\tan \theta = 49/70$, and $\theta = 35.0°$. We now get Table IV.

TABLE IV.

Frequency (MHz)	θ (degrees)	$\tan \theta$	$Z_0' \tan \theta$	Compensated Impedance
600	35.0	0.70	49	$75+j\,0$
400	23.3	0.43	30	$35-j\,5$
500	29.1	0.56	39	$55+j\,14.5$
700	40.8	0.86	60	$60-j\,12$
800	46.7	1.06	74	$37.5+j\,19$

The plot of these values, shown as the solid curve of Fig. 15.5, shows that this choice of parameters is satisfactory.

A circuit for accomplishing the compensation is shown in Fig. 15.6. Part (a) of this figure shows the input and compensating circuit, part (b) the final assembly. Referring to this figure, we see that as the signal arrives from the left, it travels through 50-Ω transmission line until plane *D-D*. Here a break in the inner conductor occurs; so that any current flowing on the inner conductor must pass through the shorted transmission line *E-F* before continuing in direction *G-H*. Line *E-F* is thus in series with the circuit.

Let us choose 1 mm as the diameter of the inner conductor of line *E-F*. The I. D. of the outer conductor of *E-F* is therefore given by the formula for Z_0 of a coaxial line, $Z_0 = 138 \log_{10} b/a$; where b and a are the outer and inner radii, respectively. For the 70-Ω line of $a = 0.5$ mm, we get $b = 1.63$ mm.

To design the 50-Ω line, we will use a 0.040-in. wall thickness for the 70-Ω line, so that the radius of the inner conductor of the 50-Ω supply line is 2.65 mm. For the 50-Ω line, for which $\log_{10} b/a = 0.362$, the inner radius of the outer conductor now becomes 6.07 mm.

It remains to assemble the unit and join it to the device to be impedance compensated. This is shown in Fig. 15.6(b).

If desired, the input 50-Ω line can be tapered into dimensions of any readily available coaxial cable. To keep the characteristic impedance at the 50-Ω level, the relation $\log_{10} b/a = 0.362(\varepsilon_r)^{1/2}$ must be maintained, where ε_r is the dielectric constant of the material filling the line.

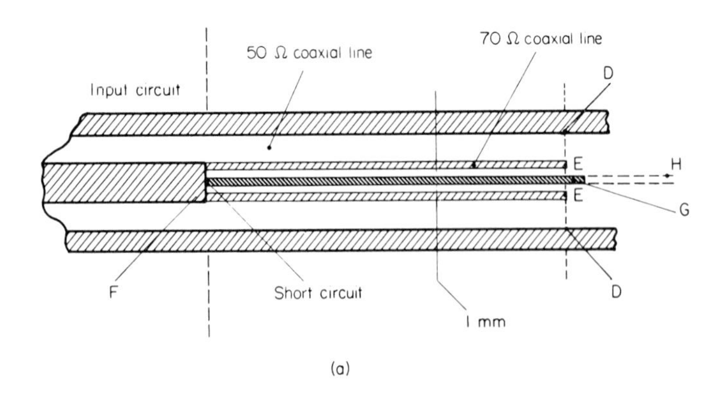

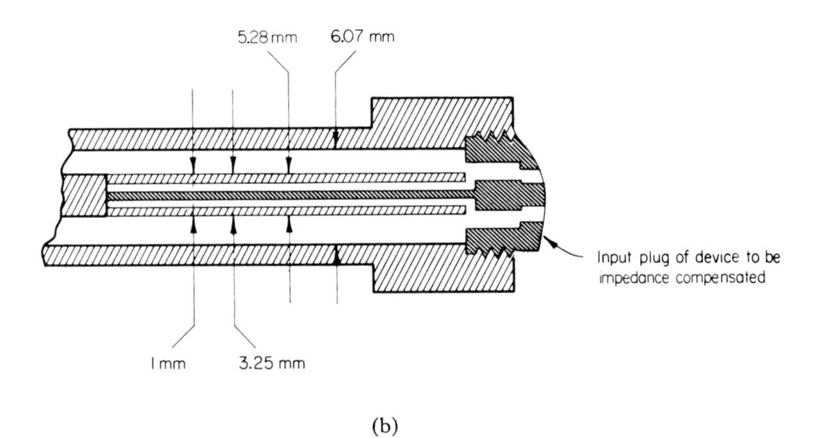

(b)

FIG. 15.6. (a) Details of the series transmission line arrangement. (b) Assembly of matching unit to input plug of device to be compensated.

15.3. Exercises

15.3.1.

Plot the locus of the normalized compensated impedance on a Smith chart. Does it fall within the required S circle?

15.3.2.

How long should the shorted line E-F be?

15.3.3.

In joining two transmission lines of the same characteristic impedance, i.e. of the same ratio of inner to outer radius, but of different absolute radii, an extra impedance is introduced. This is discussed in J. R. Whinnery, H. W. Jamison, and T. E. Robbins, Coaxial Line Discontinuities, *Proc. I. R. E.* Nov. 1944. Are there such discontinuities in the assembly of Fig. 15.6(b)? Are they serious in the frequency range in question? How could they be minimized?

15.3.4.

The assembly of Fig. 15.6(b) shows the use of a shorted air-dielectric transmission line, to which another air-dielectric line is attached at the left. In practice, the use of lines with a solid dielectric may be more advisable here. What changes must be made in the dimensions if the 50-Ω input line has Teflon as a dielectric between inner and outer conductors?

15.3.5.

The impedance compensation used here was very simple, since the real component of the impedance to be compensated was within the limits of real impedance of the definition circle. Suppose this had not been the case. What other methods of impedance compensation exist?[1]

[1] "Very High-Frequency Techniques," Vol. I by the staff of Radio Research Laboratory, Harvard Univ., McGraw-Hill, New York, 1947; or F. D. Bennett, P. D. Coleman, and A. S. Meier, The design of broad-band aircraft-antenna systems, *Proc.* I.R.E. **33,** 671, 1945.

16. BALUN AND NARROW BAND IMPEDANCE MATCHING*

16.1. Problem

It has been discovered that a gas discharge column contained in a glass cylinder can be made to radiate as an antenna when excited with signals of a frequency that is approximately $1/\sqrt{2}$ times the plasma frequency of the discharge. An investigator is given the task of devising an experimental

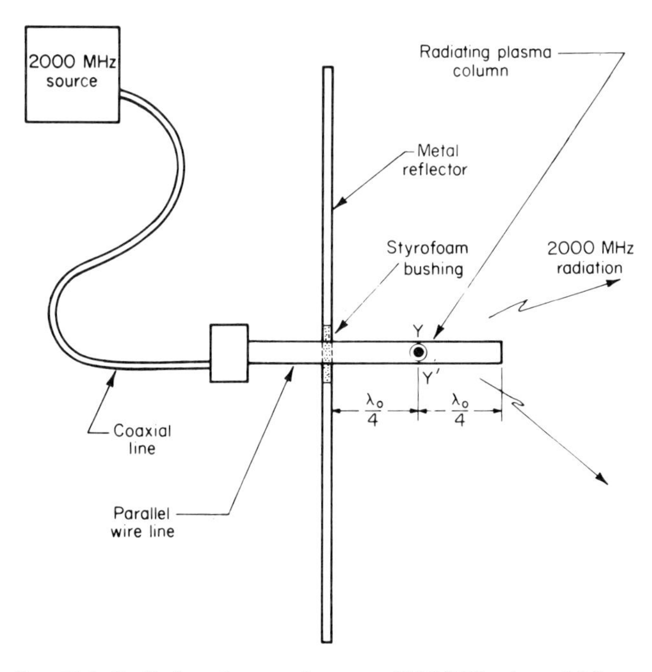

FIG. 16.1. Radiating plasma column; at 2000 MHz, $\lambda_0 = 15.0$ cm.

arrangement that allows the column to radiate as efficiently as possible when the column is supplied with signals from a 50-Ω matched source through a 50-Ω coaxial line. He is also charged with investigating the

* Problem 16 is by I. Kaufman.

radiation pattern of the radiating column, although he *is* permitted to enhance the radiation in certain directions with metallic reflectors. Eventually, he decides to use the arrangement of Fig. 16.1, in which the column is placed a quarter wavelength ($\lambda_0/4$) in front of a metal reflector. To excite the column, he uses a parallel wire line, as shown. This line is then connected to the coaxial supply line on the left side of the reflector. The parallel wire line is passed through the reflector in a hole that is appreciably larger than the wire spacing. To keep it in place, an insulating thin styrofoam bushing is wedged in the hole.

1. How should the parallel line be connected to the coaxial line?
2. According to calculations, the plasma column is expected to act like a 300-Ω load shunted across the line at plane $Y-Y'$. Design a system for transforming maximum power to this load. The frequency of operation is 2000 MHz. The outside diameter of the plasma column is 1.0 cm.

16.2. Solution

The coaxial line is an unbalanced system; the plasma column, supplied by a parallel wire line, is balanced. Direct connection of a balanced to an unbalanced system is undesirable. This is seen with reference to Fig. 16.2.

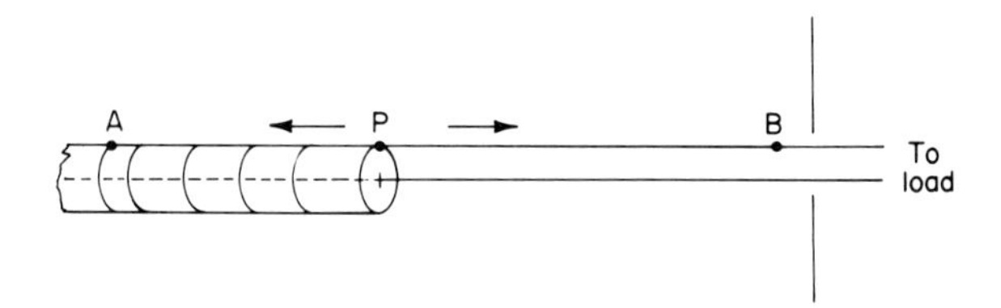

FIG. 16.2. Division of current in a system in which a parallel wire line is connected to a coaxial line.

A coaxial line cannot radiate, since there is no way for the energy to escape. A parallel wire line, while not automatically shielded, will still not radiate excessively unless the spacing between wires is an appreciable fraction of a wavelength. This is because the radiated fields from the oppositely directed currents are 180° out of phase, so that little net field remains. Should there be an inequality in the two currents, appreciable radiation *can* occur. In Fig. 16.2, the current arriving at P from the inside of the coaxial line can split into paths PA and PB. While the majority of current directed toward PB will flow to the load, that directed toward PA will return as displacement current, also eventually becoming conduction current on the center conductor at the source again. Consequently, stray radiation from the outside of the coaxial line as well as from the unbalanced parallel wire line can occur. This situation is cured by use of a balun.

There are a number of ways of constructing a balun. Since operation here is to be at a single frequency only, a narrow band balun will suffice. Consequently, we can use the simple structure of Fig. 16.3. It makes use of the impedance transforming property of a quarter-wavelength line. For such a section of line, if R is the terminating impedance, Z_0 the characteristic impedance of the line, then the input impedance is $(Z_0{}^2/R)$. In particular, if $R = 0$, then the input impedance is infinite.

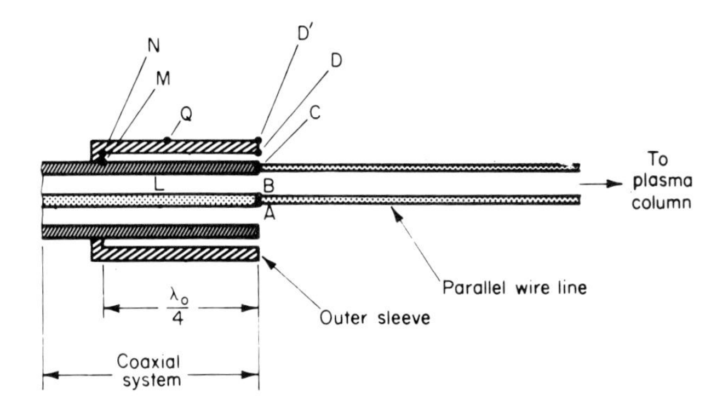

FIG. 16.3. Simple balun.

Referring to Fig. 16.3, if the outer sleeve were not present, the current flowing on the inside conductor of the coaxial line in direction LB would split at B, as before. However, with the shell now placed around the coaxial line, any current flowing from C toward M must be accompanied by an equal current returning from N to D. The equivalent circuit is therefore as in Fig. 16.4. Since any stray current flowing out of D' toward Q now has a shorted $\lambda_0/4$ line in series with it, no current can flow from C to M, nor from D' to Q. This solves the unbalance problem.

To transform maximum power to the radiating plasma column, we must provide an impedance match throughout. Let us first, therefore, design the parallel wire line for 300-Ω characteristic impedance. For the configuration of Fig. 16.5, the characteristic impedance Z_0 is $Z_0 = 276 \log_{10} 2s/d$. In addition, we have the requirement here that $(s - d) = 1.0$ cm. Simultaneous solution of these two relations yields $s = 12.0$ mm, $d = 1.96$ mm.

Assuming the plasma to be 300 Ω, and now supplied by a 300-Ω line, there should be a VSWR = 6 on the 50-Ω coaxial line. This means that only 49% of the incident power could be absorbed by the plasma radiator. To approach 100% absorption, we transform the 300-Ω impedance to the 50-Ω level with a quarter wave transformer. As stated above, for a quarter wave Z_0 line terminated in impedance R, the input impedance is $(Z_0{}^2/R)$.

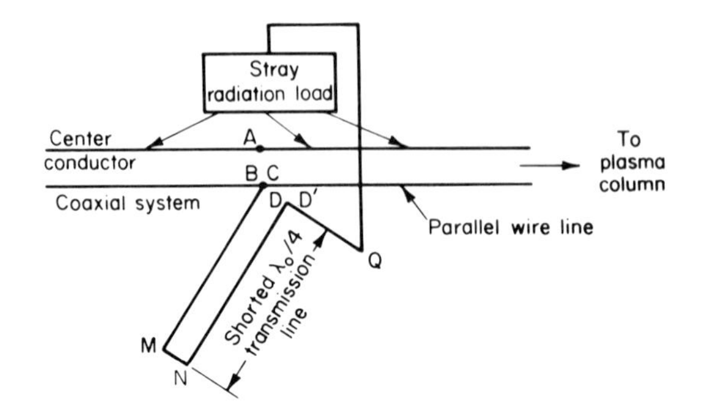

FIG. 16.4. Equivalent circuit of balun of Fig. 16.3, showing shorted quarter wave transmission line.

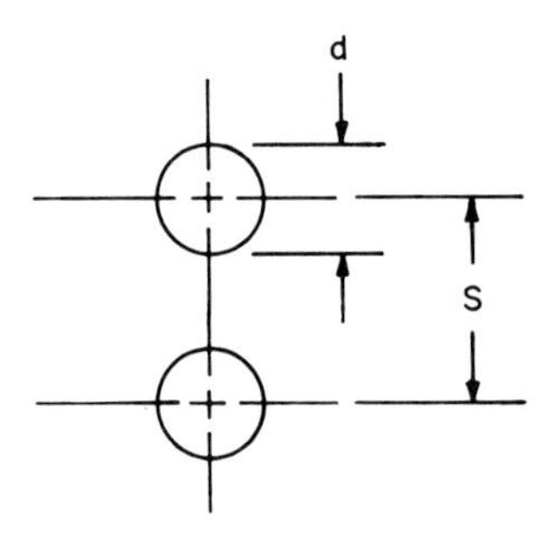

FIG. 16.5. Open wire transmission line.

In the present case, we require that $Z_0 = [(300)(50)]^{1/2} = 122\ \Omega$. Consequently, we must design a matching section of that characteristic impedance.

Because the conductors for a relatively low impedance parallel wire line ($\sim 200\ \Omega$) are of a diameter that is an appreciable fraction of the spacing, we use the more exact formula $Z_0 = 120 \cosh^{-1} s/d$. Maintaining $s = 12.0$ mm, we find here that $d = 7.62$ mm. The configuration is shown in Fig. 16.6.

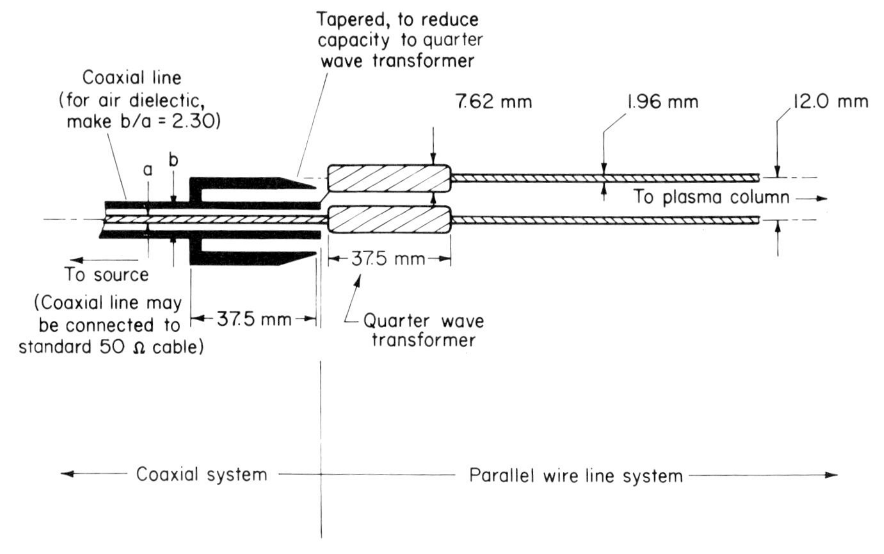

FIG. 16.6. Coaxial-to-parallel wire line system for supplying plasma column.

16.3. Exercises

16.3.1.

When the balun is fabricated, it will not be possible to fabricate it so that it will provide optimum balun action at exactly the design frequency. It must therefore be tested and modified slightly. In view of the fact that there is not supposed to be any radiation from it at the design frequency, how would you test it?

16.3.2.

After the balun has been adjusted and the plasma radiating system is operated, a standing wave ratio of VSWR = 2.5 is found to exist on the coaxial line. How would you reduce this VSWR with additional equipment?

16.3.3.

It is found that the parallel wire line sags slightly so that it must be supported. What will the supports do? Suggest ways of minimizing their effect.

16.3.4.

Suppose that instead of a plasma column, which is an impedance that varies with frequency, the line were terminated with a frequency-independent 300-Ω load. For this case, what would be the bandwidth for VSWR = 1.5 at the input to the quarter wave transformer? At the input to the balun?

16.3.5.

How would you improve the bandwidth characteristics?

16.3.6.

How would you compute the radiation from the parallel wire line? How would you reduce it?

16.3.7.

How would you measure the efficiency of radiation of the plasma column; i.e. what fraction of the power is radiated, what fraction is absorbed by the gas in the column?

16.3.8.

What is the effect of discontinuities in the transmission lines?[1]

[1] J. R. Whinnery and H. W. Jamieson, 'Equivalent circuits for discontinuities in Transmission lines,' *Proc. I.R.E.* p. 98, 1944.

17. MEASUREMENT OF PULSED MICROWAVE POWER*

17.1. Problem

It is desired to measure the amplitude and approximate pulse shape of some 27 GHz pulsed power that originates in an experiment. This power has been detected with a microwave crystal video detector. In this measurement, the output of the detector was connected through a 50-Ω cable of 1.83 m length to the terminals of the preamplifier of an oscilloscope. The oscilloscope has a frequency response that is good from dc to 30 MHz. The pulse repetition rate of the signal is 20 p.p.s. The display seen on the oscilloscope was that of Fig. 17.1.

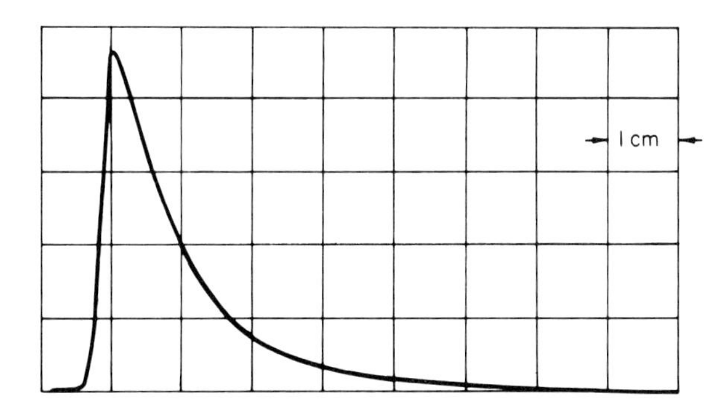

FIG. 17.1. Pulse seen on oscilloscope. Vertical sensitivity: 0.05 V/cm; horizontal sweep rate: 1.0 μsec/cm.

The experimenter knows that something is amiss, for the pulses are supposed to be only 0.5 μsec long.

Devise a technique for measuring the power level during the various portions of the pulse. You have available to you, in addition to the experimental setup already at hand, a number of other devices. Among these is a calibrated microwave receiver. The ratings of this receiver state that its bandwidth between 3dB points is 1 MHz; its sensitivity is − 80 dBm; and its image rejection is 60 dB. There are also bolometers and a bolometer

* Problem 17 is by I. Kaufman.

bridge for the 26.5–40.0-GHz range; a water load; a microwave calorimeter that can measure down to about 10 μW with accuracy $\pm$ 2% and to 3 μW with accuracy $\pm$ 5%; a klystron for 26.5–29.0 GHz with supply; microwave hardware, such as a precision variable attenuator, directional couplers, slotted line, etc.; and the usual laboratory equipment.

17.2. Solution

It would have been very pleasant to use the calibrated microwave receiver and just connect its output to an oscilloscope. However, its bandwidth is only $\pm$ 0.5 MHz. Since it is required to observe some detail in pulses of duration 0.5 μsec, bandwidths of better than 5 MHz are required. The receiver can therefore not be used.

Standard bolometers for the frequency range to be measured have response times in the 100-μsec range; so they are obviously useless for a direct measurement. The calorimeter can be expected to have a response time of several minutes. It can therefore be used only to measure average power, not instantaneous or peak power.

On the other hand, even the crystal detector appears to have a response time of 2 μsec, according to Fig. 17.1. Fortunately, it will be found that this is not really the case. The 2-μsec response is not inherent to the crystal, but to the circuit in which it is used. While the microwave equivalent circuit of the crystal, as used here, is chiefly a nonlinear resistance, in the video range it behaves more like a diode. This is seen as the crystal output is "loaded" by placing resistances of various magnitudes across the oscilloscope input to which the crystal output is connected. A typical situation is shown in Fig. 17.2. It is seen that an accurate representation of the pulse waveform is obtained with the circuit used only if the circuit is loaded with 100 Ω or less.

Of all the devices available for power measurement, therefore, only the crystal, properly loaded, can be used here. However, since its indication is only relative, it must be calibrated. Before calibrating the crystal, it is best to place it in the actual circuit whose power is to be measured. By use of any tuners that may be a part of the crystal mount, and by rotating the crystal in its socket, it should be adjusted for maximum voltage output. Once these adjustments have been made, they must not be changed during the entire calibrating procedure.

An important point is that the frequency of the pulse must be measured so that the calibrating circuit can be set to this frequency. Finally, the amplitude of the detected pulse should be noted.

We now place the crystal assembly into the calibration circuit, such as the one shown in Fig. 17.3, being very careful to always "ground" the unit before making connection with the center pin of the connector that is the crystal connection, and always starting with the attenuator adjusted for maximum attenuation. As indicated above, the resistor that loads the crystal must be in place during the measurement, as well as during the

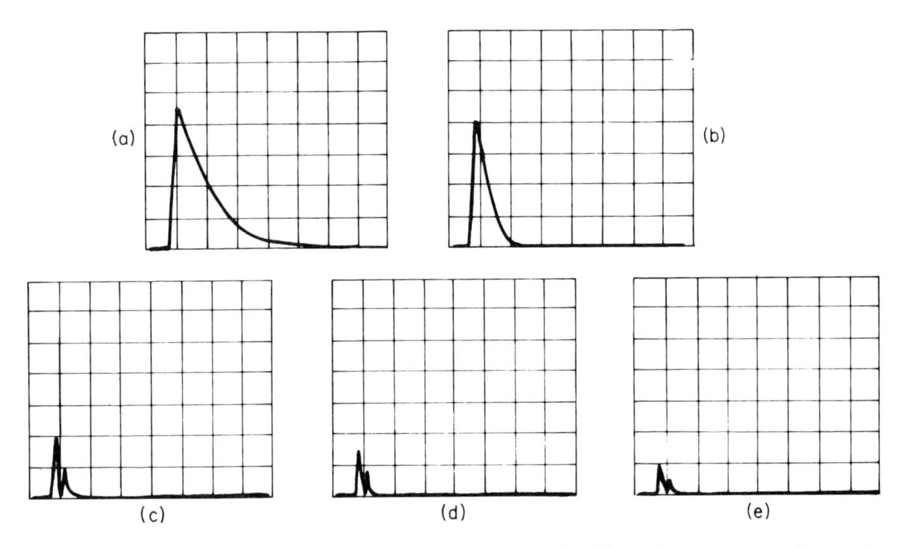

FIG. 17.2. Oscilloscope traces of pulse to be measured, with various amounts of crystal loading: (a) no loading; (b) 1500 Ω: (c) 200 Ω; (d) 100 Ω; (e) 50 Ω. Horizontal sweep rate is 1.0 μsec/cm.

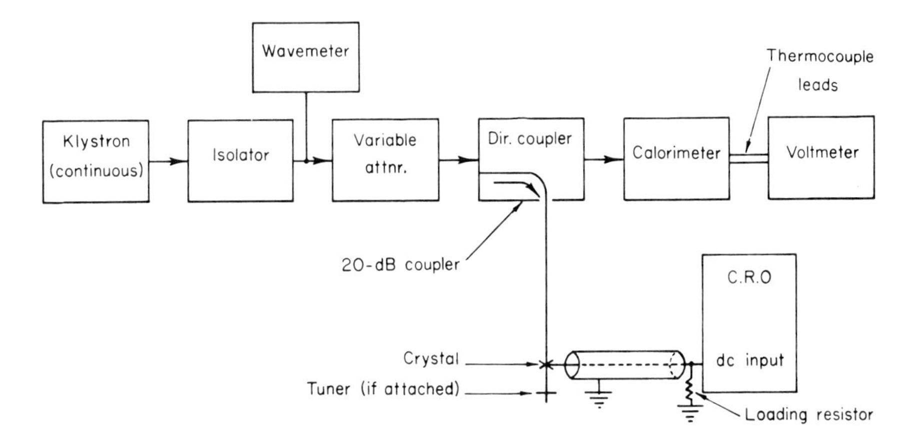

FIG. 17.3. Crystal calibration circuit.

calibration. In the circuit of Fig. 17.3, the calorimeter may be replaced by a matched bolometer.

The klystron is set to the proper frequency. The directional coupler must have been previously calibrated. By means of the variable attenuator, the power level in the main line is now adjusted to the level that produces the same amount of deflection on the oscilloscope (CRO) that the pulse to be measured produced. The amount of power corresponding to this level is the power measured by the calorimeter, minus the coupling coefficient of the directional coupler.

After calibration, the earlier measurement should be repeated, to ensure that the crystal characteristics have not changed.

17.3. Exercises

17.3.1.

On the basis of Fig. 17.2, draw an approximate equivalent video circuit for the crystal detector.

17.3.2.

Suppose the sensitivity of the crystal, loaded with sufficient resistance to reproduce the pulse shape, is just barely insufficient to get a good display on the oscilloscope. In looking for methods of amplifying the pulse, you find an amplifier with a voltage gain of 100; with a 3-dB frequency response from 3 Hz to 25 kHz and a drop-off rate of 18 dB/octave below 1 Hz and above 50 kHz. Is this amplifier of use to you, and why?

17.3.3.

Suppose the sensitivity of the crystal loaded with 100 Ω is insufficient to get an accurate representation of the pulse shape. However, the signal level with 200 Ω of loading is sufficiently high to get a good indication of the peak signal level and some of the pulse structure. Suggest ways of improving your measurement circuit so that you can see the details of the pulse.

17.3.4.

At low levels of video detection, the output voltage of the crystal is proportional to microwave power, not microwave field intensity. Why? Suggest an experimental setup that ensures that the level at which the crystal has been calibrated is not in the "saturated" range, where the proportionality no longer holds. Why is it undesirable to operate in this "saturated" range, even for single point calibration?

17.3.5.

A directional coupler is shown in Fig. 17.3. Suppose that this coupler is of the crossed-guide variety, which allows simultaneous sampling of incident and reflected power. What precaution must you use in connecting the coupler in this circuit?

17.3.6.

Suppose the calorimeter of Fig. 17.3 is not available, but a bolometer *is* available. The bolometer fits into a standard crystal mount, which has a tuning plunger following the bolometer. It is found that this tuning plunger can change the power reading of the bolometer by orders of magnitude. What measurement should you perform to ensure that your reading of bolometer power in the calibration procedure is correct, or how it can be corrected?

17.3.7.

The calibration procedure outlined above prescribes the use of the "dc input" of the oscilloscope. You find, however, that on the low range required for the measurement the oscilloscope will only amplify from 3 Hz to 25 MHz. Fortunately, the klystron supply that you possess has provision for the following modes of operation: continuous; 1000 Hz square wave; pulse; and 60 Hz sine wave. How will this help you?

18. MICROWAVE CAVITY RESONATOR*

18.1. Problem

A microwave cavity is to be built so that a pulsed 10-kV electron beam, which will sweep through it from one end to the other in the manner of an oscilloscope sweep, can be shot through it. The cavity is to have the configuration of Fig. 18.1. Here the electrons will enter essentially perpendicular to the top face.

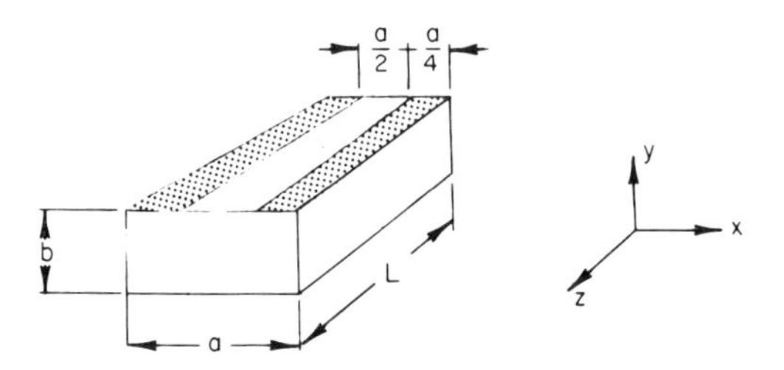

FIG. 18.1. Microwave cavity; $a = 7.12$ mm.

The cavity is to be resonant to approximately 34 GHz and is to have high Q. It should be slightly tunable in frequency. The height b is to be such that the electron transit angle is $1\frac{1}{2}$ cycles of the 34-GHz radiation. The length L is to be $1\frac{1}{2}$ guide wavelengths; so that the cavity oscillates in the TE_{103} mode. (The E field has only a y component.)

1. Determine the cavity dimensions.
2. Estimate the Q on the basis of solid walls (i.e. no electron passage).
3. What expected difficulties do you foresee arising from the need for beam passage, and how can they be remedied?
4. Suggest a method of tuning the cavity.
5. Output power is to be delivered to RG-96/U waveguide. How can this be accomplished? (Note: The power is generated by the electron beam passing through the cavity.)

* Problem 18 is by I. Kaufman.

18.2. Solution

1. The height b is found from the requirement of a transit angle of $1\frac{1}{2}$ cycles of a 10-keV electron. The velocity of a 10-keV electron is found from $\frac{1}{2}mv^2 = eV$, where m is the electron mass which equals 9.11×10^{-31} kg, e is the electron charge which equals 1.60×10^{-19} C, v is the electron velocity, and V is the beam voltage (here, 10^4 V). For 10-keV electrons, $V = 5.33 \times 10^7$ m/sec. For 34 GHz, $1\frac{1}{2}$ cycles $= (1.5)\ 1/[(34)\ (10^9)]$ sec. The dimension b is therefore

$$b = (1.5)\left[\frac{1}{(34)\ (10^9)}\right][5.93 \times 10^7] = 2.62 \times 10^{-3} \text{ m.}$$

To find the cavity length, we must first find the guide wavelength. The free-space wavelength, λ_0, is $(3)\ (10^8)/(34)\ (10^9) = (8.83)\ (10^{-2})$ m. The cutoff wavelength, λ_c, is $2a = (1.422)\ (10^{-2})$ m (see Fig. 18.1). The guide wavelength, λ_g, is given by $\lambda_g = \lambda_0/(1 - (\lambda_0/\lambda_c)^2)^{1/2}$. The cavity length here is therefore

$$\tfrac{3}{2}\lambda_g = (1.693)\ (10^{-3}) \text{ m} = (0.663'').$$

The inside dimensions of the cavity are therefore:

$$a = 7.12 \text{ mm}$$

$$b = 2.62 \text{ mm}$$

$$L = 1.69 \text{ mm.}$$

2. To estimate the Q, we recall that the cavity Q is given by $Q = (\omega U)/P$, where ω is the angular frequency of resonance, U is the stored energy, and P is the power lost in walls when the energy stored has the value U. To find these quantities, we refer to a text or reference on microwave cavities. For example, a good discussion of the box resonator is given in S. Ramo and J. R. Whinnery, *Fields and Waves in Modern Radio*, Wiley, N.Y., 1944; pp. 383–389. Here it is shown that for a box resonator, when the peak electric field is E_0, the stored energy U for a cavity of length $\lambda_g/2$ is

$$U_{\lambda g/2} = \left(\frac{1}{8}\right)\varepsilon_0\, ab\left(\frac{\lambda_g}{2}\right)E_0 \quad \text{where} \quad \varepsilon_0 = 8.854 \times 10^{-12}.$$

The power loss in the walls is

$$P = \frac{R_s\,\lambda_0{}^2}{8\eta_0{}^2}E_0{}^2\left[\frac{(b)\ (\lambda_g/2)}{a^2} + \frac{ab}{(\lambda_g/2)^2} + \frac{a}{(2)\ (\lambda_g/2)} + \frac{(\lambda_g/2)}{2a}\right].$$

The terms in the square brackets, above, are due to power loss in the various walls. Referring to Fig. 18.2, we have the following tabulation.

		Loss in walls
First term–	–	1 and 2
Second term	–	3 and 4
Third term	–	5 and 6
Fourth term	–	5 and 6

As before, λ_0 is the free-space wavelength; $\eta_0 = 377\ \Omega$; R_s is the surface resistivity. MKS units are assumed.

Our cavity is $1\frac{1}{2}$-wavelengths long. Therefore, for a given E_0, our total stored energy is three times that of the corresponding cavity of length $\lambda_g/2$. However, the dissipated power is less than three times that of the

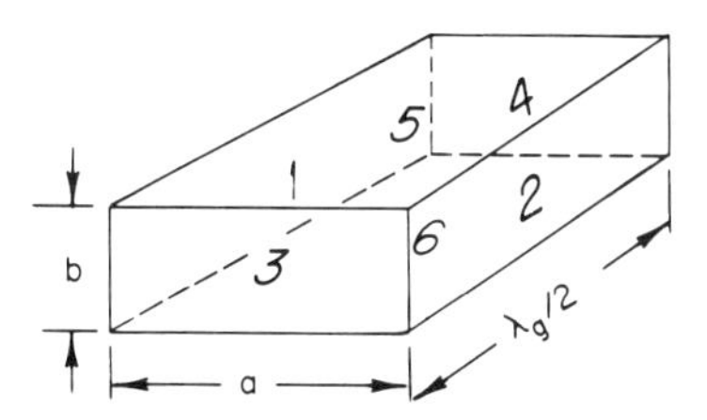

FIG. 18.2. Half-wavelength rectangular box resonator.

$\lambda_g/2$ cavity, since there is no end wall dissipation for the center section and for half of each end section. Consequently, we have

$$P = \frac{R_s \lambda_0^2}{8\eta_0^2} E_0^2 \left[3\,\frac{(b)(\lambda_g/2)}{a^2} + \frac{ab}{(\lambda_g/2)^2} + 3\,\frac{a}{(2)(\lambda_g/2)} + 3\,\frac{(\lambda_g/2)}{2a} \right],$$

$$U = 3\,\frac{\varepsilon_0 (ab)(\lambda_g/2)}{8} E_0^2.$$

Since the cavity Q is to be high, and since an electron beam is involved, we choose a high conductivity material that is nonmagnetic. The preferred material here is copper, particularly of the oxygen-free variety (to allow brazing without flaws). The surface resistivity of copper is given by $R_s = (2.61)\,(10^{-2})\,\sqrt{f}$, where f is the frequency. (See S. Ramo and J. R. Whinnery, op. cit.; p. 210.) For 34 GHz therefore, $R_s = (4.80)\,(10^{-2})$. Substitution of the other parameters yields

$$U = (3.49)\,(10^{-19})\,E_0^2$$
$$P = (1.50)\,(10^{-11})\,E_0^2.$$

The cavity Q is therefore 4980.

3. If the top and bottom walls have large apertures, to allow for beam passage without loss, a small amount of 34-Gc power will be radiated from the cavity. This leakage of power can easily reduce the Q to an extremely low value and make the cavity useless. Consequently, it must be minimized.

In general, radiation will emanate from an aperture if this aperture interrupts lines of current that would flow if the aperture had not been present (if the wall were solid). The linear current density $\mathbf{J}$ at a wall is given by $\mathbf{J} = \mathbf{n} \times \mathbf{H}$, where $\mathbf{n}$ is the unit normal, $\mathbf{H}$ is the magnetic field intensity. It therefore behooves us to examine the direction of magnetic field lines at top and bottom walls of a solid cavity. The pattern of magnetic field lines and corresponding current lines (found from the equation, above) is given in Fig. 18.3. It is seen that a long, narrow slot at the center

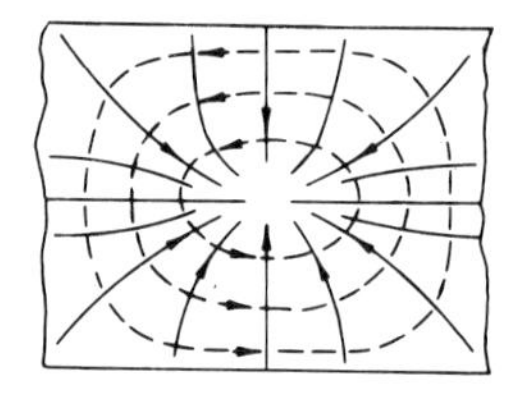

Fig. 18.3. Magnetic field configuration (dashed lines) and current distribution (solid lines) in top wall of center section of $1\frac{1}{2}\ \lambda_g$-length cavity.

will not interrupt current lines. As soon as the slot is widened, however, so that an electron beam of appreciable dimensions can pass through the cavity, current lines are interrupted. The standard method of allowing electron beam passage without appreciable diminution of Q is to cover the slot by a screen or a grid. Several types of screen are available. The reader is referred to texts on electron tube technology for details.

For a device of low duty cycle, a tungsten wire mesh of 25.4 μm wires, spaced 254 μm by 254 μm, will be satisfactory. The tungsten should be gold plated, for low secondary electron emissitivity, and to allow brazing to the edge of the slot.

4. To tune the cavity, we recall that a resonant circuit "resonates" at the frequency at which the inductive stored energy exactly equals the capacitive stored energy. A microwave cavity obeys the same principle. Consequently, to detune a cavity from resonance at a particular frequency, it is necessary to change the balance of energy. This is done by changing either the volume available for inductive energy or by changing the spacing between walls in which capacitive energy is stored. A new resonant frequency results, at which the two energies are again equal.

In circuit terms, the resonant frequency is given by $\omega_0 = 1/(LC)^{\frac{1}{2}}$, where L is the inductance and C is the capacitance. A decrease in either L or C will raise the resonant frequency and vice versa.

A simple way of reducing the volume for inductive energy is to insert metal rods into the region of magnetic fields. Use of this idea is made in Fig. 18.4.

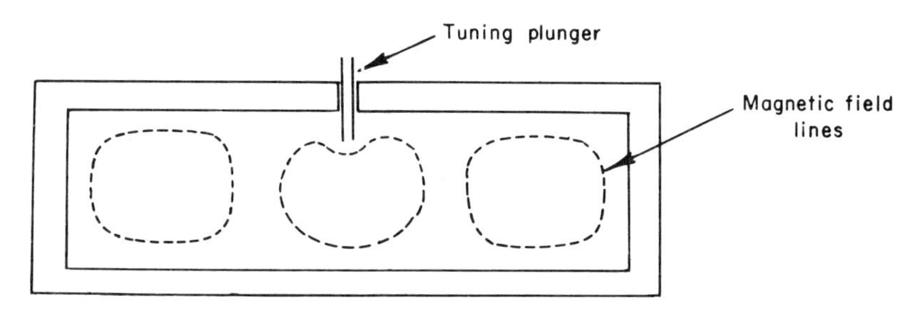

FIG. 18.4. Method of inductive tuning of cavity. To maintain symmetry, three tuning rods can be used—one for each $\lambda_g/2$ region.

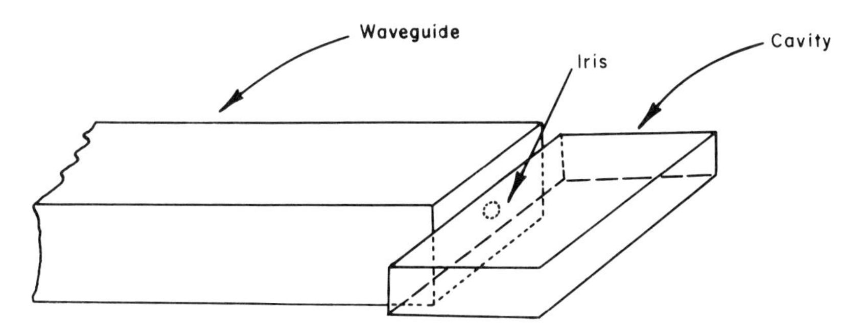

FIG. 18.5. Method of coupling to cavity with waveguide.

5. While coupling to cavities at the lower frequencies is frequently performed with magnetic loops or electric probes, the common method at high microwave frequencies is iris coupling. This makes use of the fact that radiation outward from the cavity occurs wherever current lines are interrupted, and that radiation inward is possible wherever outward radiation can occur.

Examination of the lines of current flow will show that coupling will occur to our cavity if a waveguide is connected to our cavity through an iris in either end or side walls. A scheme of coupling through the side wall is shown in Fig. 18.5. (The procedure for determining the size of the iris is treated in a separate problem.)

18.3. Exercises

18.3.1.

A computation of the amount of 34-GHz power generated was made on the basis that no electrons intercept the screen at top and bottom of the cavity, yet that the cavity Q is undiminished from the value computed for solid walls. In practice, it is found that the screen wire chosen is indeed nearly as good in determining Q as a solid wall. How much will the power be reduced because of electron interception?

18.3.2.

Each of the screen wire apertures can be considered as a rectangular waveguide of height and width 229 μm and length 25.4 μm. How much attenuation (in decibels) occurs in such a waveguide at 34 GHz?

18.3.3.

In testing the cavity (with waveguide connected), it is found to be resonant at 33.85 GHz. This is presumed to be the resonance sought. How can you be sure that the correct mode is excited?

18.3.4.

The tuning plunger of Fig. 18.4 is found to lower the Q. One reason could be the leakage of energy through the coaxial line formed by the plunger and the hole in the side wall through which the plunger slides. This assumes that the plunger does not make good contact with the metal —a reasonable assumption. How could you prevent such leakage of energy?

18.3.5.

In the operation of the plunger of Fig. 18.4, it is found that as the plunger is inserted into the cavity more and more, the resonant frequency is at first raised, then lowered. Why?

19. COUPLING OF A MICROWAVE CAVITY TO A WAVEGUIDE*

19.1. Problem

A microwave cavity, designed to be resonant at 34 GHz, is to terminate a TE_{10}-mode rectangular waveguide. It is to be coupled to this guide by an iris in a thin wall separating the two. When supplied with power at resonance through this waveguide, 40% of the incident power is to be absorbed in the cavity, 60% is to be reflected. At this degree of coupling, the maximum frequency selectivity possible is to be maintained.

Design an experiment for measuring the resonant frequency and unloaded Q of the cavity, and for selecting the correct iris size for the 40% coupling. The expected value of unloaded Q for this cavity is about 5000.

19.2. Solution

There are a number of different procedures that can be used for solving this problem. We discuss here a particularly simple method that holds for the thin wall case, in which negligible coupling reactance is introduced. It follows the treatment by R. A. Lebowitz, *IRE Trans. Microwave Theory Tech.* **MTT4**, 51 (1956).

We require a tunable oscillator, such as a reflex klystron, a slotted line, a directional coupler (20 dB), an attenuator, and an oscilloscope. The circuit is shown in Fig. 19.1. If the 20-dB attenuator shown in this circuit can be replaced by a circulator, more signal will be available, so that less care in the prevention of "hum" signals arising from ground "loops" will be required.

We will assume, here, that the generator available for the experiment is a reflex klystron, which can be tuned mechanically over a considerable frequency range and tuned electronically (by repeller modulation) over a small range.

The initial step is to find the resonant frequency. With no iris, i.e. with a solid wall between waveguide and cavity, nearly all of the energy incident

* Problem 19 is by I. Kaufman.

on the wall is reflected. Almost 100% reflection occurs also with an iris if the frequency of the incident wave is not equal to or near the resonant frequency of the cavity. Consequently, to find the resonant frequency, we drill a very small hole in the thin wall separating waveguide from cavity, mechanically tune the klystron through the range of expected resonance

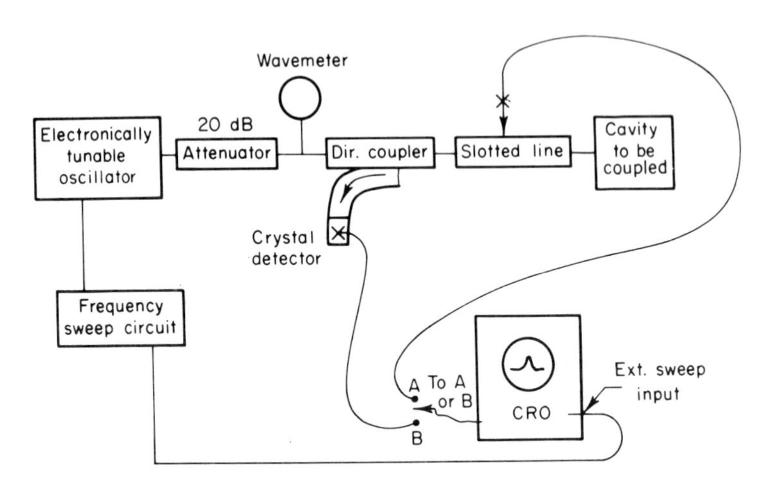

FIG. 19.1. Circuit for performing cavity coupling measurements.

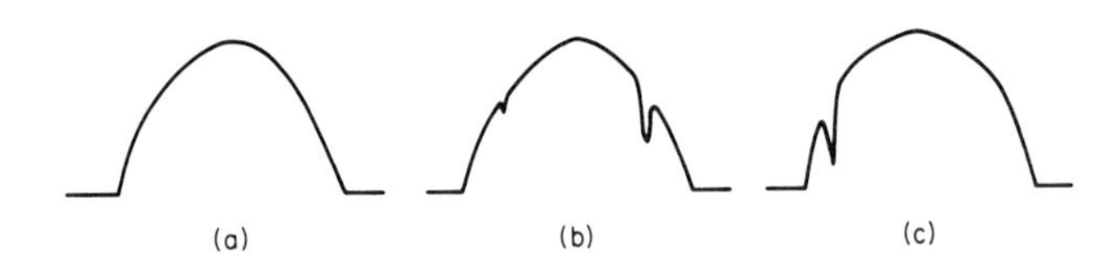

FIG. 19.2. Reflected power, as seen at the output of the detector following directional coupler: (a) with neither wavemeter nor cavity resonance within the klystron mode; (b) with cavity resonance dip on left and wavemeter dip on right; (c) with the two resonances superimposed. Since the oscilloscope is set for "exterior sweep input," the horizontal coordinate is frequency, although the scale is usually not linear.

while the output frequency is being swept electronically by repeller modulation, and observe the pattern of reflected power at the output of the directional coupler. When the resonance is found, it will be observed as in Fig. 19.2. By superimposing the absorption dip of the wavemeter on the dip of the cavity, the resonant frequency is found from the wavemeter calibration. (In a more exact technique, another attenuator is inserted between wavemeter and directional coupler, so that the two resonances will not interfere cooperatively and shift the resonance slightly.)

It is possible that no cavity absorption dip will be observed on the first try. The coupling hole is then enlarged and the process repeated.

To adjust the coupling iris for 40% power absorption, we simply enlarge it progressively until the reflected power at the resonance dip drops to 60% of the value it would have if there were no iris (see Fig. 19.3).

In all of the measurements of relative power and relative fields, it must be remembered that when a crystal detector is used as a video detector, as in Fig. 19.1, it is a square-law device. Consequently, the vertical scale of the oscilloscope display is in units of relative radio-frequency power, or relative units of $(E_{\mathrm{rf}})^2$. To ensure linearity in these units, the power incident on the crystal should not exceed a few microwatts.

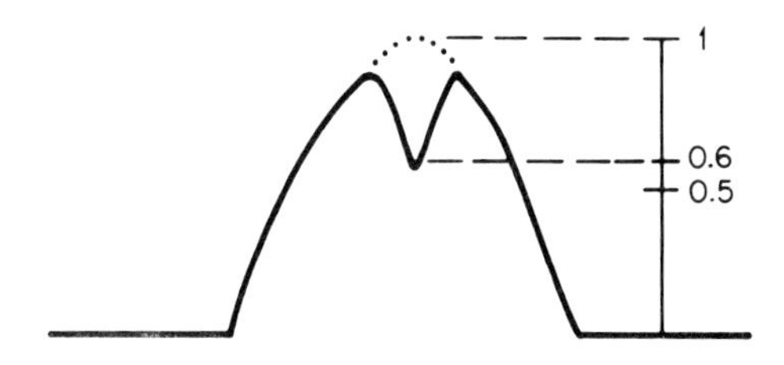

FIG. 19.3. Reflected power trace for 40% power absorption in cavity.

Since 40% absorption of incident power corresponds to a VSWR of 7.87, it would appear that we could also use the slotted line at the fixed frequency of resonance of the cavity and enlarge the coupling iris until the VSWR is 7.87. It will be found in the laboratory, however, that this technique is not feasible unless the signal source is highly stabilized in frequency by such methods as discriminator circuits, etc., for otherwise the frequency drift is too large.

On the other hand, if a swept frequency source is used, the use of a slotted line for measurement of the percentage of power absorption becomes possible again. Furthermore, this technique will also enable us to measure the cavity Q. We will discuss the method of power measurement and from there lead into the scheme of measurement of unloaded Q.

When a cavity is excited from a source whose frequency is the frequency of cavity resonance, the peak energy stored in magnetic fields exactly equals that stored in electric fields. There is thus no "reactive power" supplied to the cavity, only real power, which is dissipated in the cavity walls. As soon as the signal source is detuned from resonance, however, large amounts of reactive power are required. Accordingly, the cavity behaves exactly as a lumped-constants resonant circuit. While it would appear that this situation is true only at the iris plane, because of the repetitive nature of a transmission line, this behavior is observed at a number of planes in the waveguide feeding the cavity. Specifically, there are two sets of such planes, referred to as the planes of the "detuned open" and "detuned short," which are separated by quarters of a guide wavelength from each other. The equivalent circuits at these two sets of planes are shown in Fig. 19.4. The

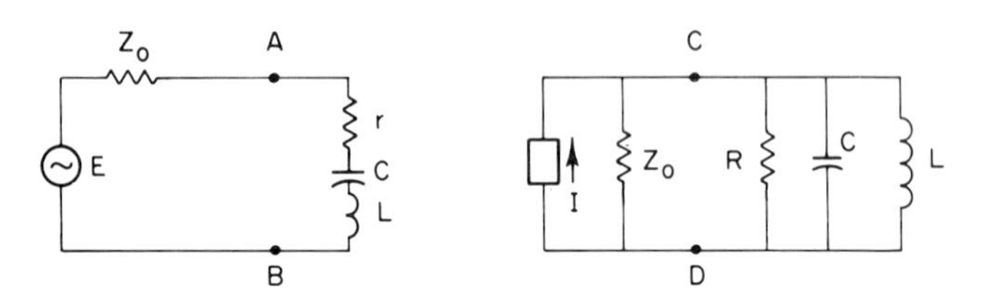

FIG. 19.4. Equivalent circuits of waveguide-cavity system at (a) detuned open and (b) detuned short. E and I here are taken to be RMS values. It is assumed that the output impedance of the source is also the waveguide characteristic impedance Z_0.

FIG. 19.5. Power detected by probe of slotted line at detuned open when cavity resonance is at center of klystron mode.

names "detuned open" and "detuned short" are used because of the equivalent impedance that the cavity presents when the signal frequency differs appreciably from the resonance frequency.

If we have chosen a slotted line (Fig. 19.1) that is at least one half-guide wavelength long, at least one of each of the two sets of planes will be found along the line.

Since the impedance at the detuned open when not at resonance is essentially an open circuit, then in this "off" resonance condition this plane is a position of maximum electric field in the standing wave pattern. When "on" resonance, however, the cavity absorbs some energy. As the klystron is swept through resonance, the output of the traveling probe located at a detuned open is therefore as shown in Fig. 19.5. In fact, the position of a detuned open is found by moving the probe until a symmetric dip is obtained. It will be noted that an increase of the coupling iris is accompanied by a deepening of the dip. Referring to Fig. 19.4(a), we conclude that an increase in the coupling corresponds to a reduction in r in the equivalent circuit.

We are now interested in the "incident power" in terms of the circuit parameters of Fig. 19.4(a). Since this is the power that would be absorbed by an infinite length of line or a matched load, it is obviously the quantity $E^2/4Z_0$.

If the power absorbed by the cavity is to be only 40% of the incident power, we find that r is given by

$$\frac{(0.40E^2)}{4Z_0} = \frac{(Er/[r + Z_0])^2}{r},$$

so that we have either

$$r = 0.127Z_0 \qquad \text{when} \quad V_{AB} = 0.888E,$$

or

$$r = 7.87Z_0 \qquad \text{when} \quad V_{AB} = 0.113E.$$

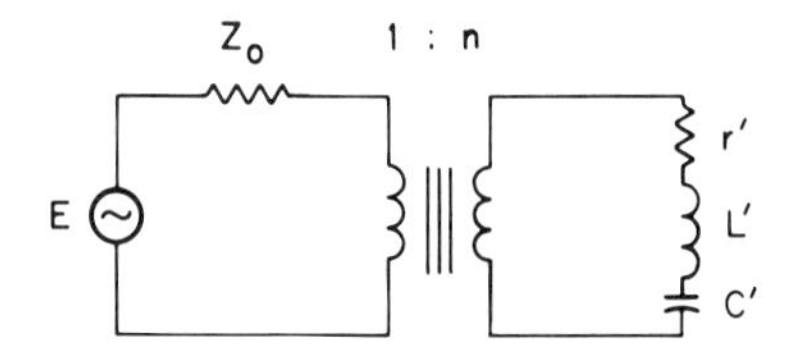

FIG. 19.6. More complete equivalent circuit at detuned open.

Accordingly, there are two iris diameters that will satisfy the 40–60% condition. In each case, of course, the VSWR is 7.87.

It was required that of the two iris sizes we choose the one that results in the maximum amount of frequency discrimination. We therefore recall that the iris is actually a transformer; so that a more complete equivalent circuit at the detuned open is that of Fig. 19.6. Here, as the iris is altered, r', C', and L' remain fixed, but n changes. The loaded Q, Q_L, which is a measure of frequency discrimination of the circuit, as calculated by using impedances of the "cavity" side of the transformer, is now $Q_L = \omega_0 L'/(R_S' + n^2 Z_0)$. As n^2 increases, Q_L decreases. For maximum frequency discrimination, Q_L should be as large as possible so that n should be as small as possible. Since at resonance $Z_{AB} = r = r'/n^2$, maximum Q_L is obtained when r is as large as possible. Consequently, the iris is to be opened only to the smaller diameter, when

$$V_{AB} = 0.888E.$$

As stated above, the output voltage of the detector is proportional to the square of the rf voltage. Consequently, the correct iris size is that corresponding to a detected voltage at the notch of Fig. 19.5 of 0.79 of that which would exist if the resonance were missing.

The final requirement of this problem is to measure the unloaded Q, Q_0. This involves a measurement of the frequency response. In the series-resonant circuit which exists at the detuned open, this could only be obtained by a measurement of the relative current response. The probe of a slotted line does not respond to waveguide current, however, but waveguide electric field. We must therefore move to the detuned short, where the equivalent circuit of Fig. 19.4(b) holds. Here the frequency response *is* given by a voltage response, which will be detected by the slotted line probe. As seen from this circuit, the frequency response is a loaded Q (Q_L) response, for it is determined by the parallel combination of R and Z_0.

If Δf is the frequency deviation from resonance at which the voltage V_{CD} drops to 0.707 of its value at resonance, and if f_0 is the resonant frequency, then

$$Q_L = \frac{f_0}{2\,\Delta f}\,.$$

19.3. Exercises

19.3.1.

Sketch the curve of detected voltage versus frequency in the measurement of Q_L. Where are the 0.707 points located on this curve? (Recall the method of response of the detector.)

19.3.2.

In the equivalent circuit of Fig. 19.4(b), the unloaded Q, Q_0, is $Q_0 = R/2\pi f_0 L$. How is Q_0 determined from Q_L?

19.3.3.

In the discussion above, we assumed that the source impedance was equal to the waveguide characteristic impedance. Is this of importance?

19.3 4.

Sketch the reflected power curves that result as the iris is progressively opened from a very small to a very large diameter. Sketch the corresponding curves seen at the detuned open and detuned short.

19.3.5.

Why is it important in these measurements to isolate the klystron by either a 20-dB attenuator or a circulator?

19.3.6.

Suppose the cavity has been adjusted for 40% coupling. It is now desired to "match" the cavity to the guide at the resonant frequency. How can this be accomplished without changing the iris diameter, but with external components?

19.3.7.

If the directional coupler available for the measurements is a crossed-guide type, which allows simultaneous observation of incident and reflected power, what precaution must be used to ensure correct measurements?

19.3.8.

What test must be made to ensure square-law behavior of the detecting crystals? It is stated that 40% absorption of the incident power corresponds to a VSWR of 7.87. Show this.

19.3.9.

Suppose you had a slotted line with loop coupling, instead of the usual probe coupling. How would this change the displays?

19.3.10.

What could be the effect of inserting the probe of the slotted line too deeply into the waveguide?

19.3.11.

A cavity is to be used as a transmission cavity, i.e. as a filter inserted in a waveguide system. It is to be coupled to the two sides in equal amounts. If the unloaded Q of the cavity is 5000 and the loaded Q is to be 1000, how much is the signal attenuated in the cavity? Express the answer in decibels.

19.3.12.

What procedure would you use for getting the correct iris sizes for the cavity of Exercise 19.3.11?

20. PHASE SHIFTER*

20.1. Problem

In a measurement circuit, shown in Fig. 20.1, a microwave source supplies two signal paths through a power splitter. Power passes in one of the signal paths through a test section to the left side arm of a magic tee.

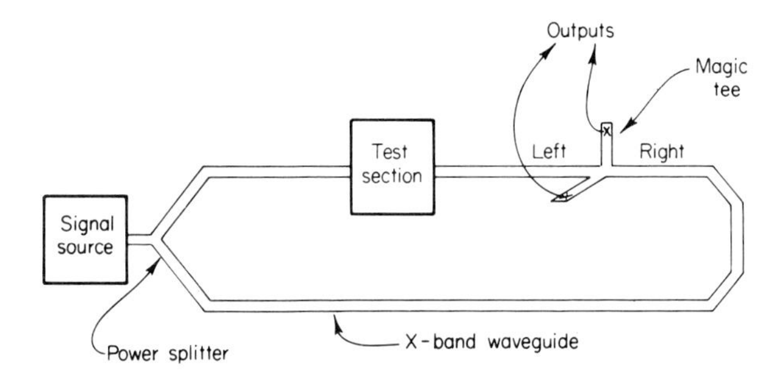

FIG. 20.1. Test circuit requiring method for adjustment of amplitude and phase.

It passes directly to the right side arm of the magic tee in the other path. In the operation of this circuit, it is found necessary to provide a continuous manual adjustment of the amplitude and phase of the signal passing to the right side arm. A variable attenuator is available, but a phase shifter is not.

Design a simple phase shifter for this job. The frequency of operation is 9.0 GHz.

20.2. Solution

A number of solutions are possible. The simplest scheme is the use of a squeeze section. This device makes use of the fact that at a constant frequency, the guide wavelength of a waveguide is a function of the guide dimensions.

* Problem 20 is by I. Kaufman.

In a rectangular waveguide operating in the TE_{10} mode, the guide wavelength λ_g is

$$\lambda_g = \frac{\lambda_0}{[1 - (\lambda_0/2a)^2]^{1/2}},$$

where λ_0 is the free space wavelength and a is the waveguide width (see Fig. 20.2). Consequently, by varying a, λ_g is changed; thereby the phase shift per unit length is altered.

We consider standard X-band waveguide, with dimension $a = 22.86$ mm. At 9.0 GHz, $\lambda_0 = 3.331$ cm, $\lambda_g = 4.862$ cm. Therefore, for a length L of this guide, a traveling wave is shifted in phase by

$$(360°)\frac{L_{cm}}{4.862}.$$

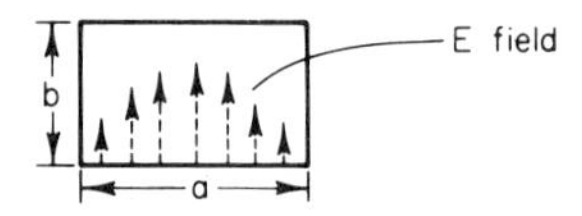

FIG. 20.2. Rectangular waveguide in TE_{10} mode.

By "squeezing" the guide, i.e. reducing a, the guide wavelength is increased, so that there is now less phase shift per unit length than before. The difference in phase is

$$360°\, L_{cm}\left[\frac{1}{4.862} - \frac{1}{\lambda_{g2}}\right]$$

where λ_{g2} is the new guide wavelength.

For complete coverage, we want this difference to be at least 360°. Consequently, the result is

$$\frac{1}{L_{cm}} = \frac{1}{4.862} - \frac{1}{\lambda_{g2}}.$$

If we arbitrarily choose $L = 20$ cm, we find that $\lambda_{g2} = 6.43$ cm. From the formula for λ_g we get $a = 1.95$ cm. Thus, if the guide width can be varied manually from 2.286 to 1.95 cm, a change of 360° in phase shift can be achieved.

It will therefore be necessary to use a section of waveguide that is considerably longer than 20 cm; slot it on top and bottom, and attach a mechanical fixture for squeezing the sides together and forcing them apart.

The scheme is shown in Fig. 20.3. The actual mechanical arrangement can take many forms.

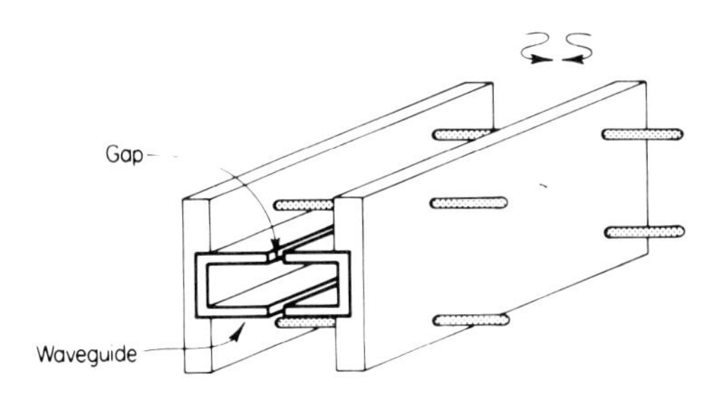

FIG. 20.3. "Squeeze section" phase shifter.

20.3. Exercises

20.3.1.

How would you test to see how much differential phase shift can be achieved?

20.3.2.

Another simple type of phase shifter uses a dielectric vane. Design such a phase shifter.

20.3.3.

A magic tee with coupled "tunable shorts" in the side arms can also be used as a phase shifter. Describe how.

20.3.4.

It is desired to use a varactor diode for accomplishing an electronically variable phase shift. Such a diode acts as a variable capacitor, whose capacity is a function of the applied dc bias voltage. Devise a scheme, using either a magic tee or a circulator, for construction of such a phase shifter.

21. WAVEGUIDE INTERFEROMETER*

21.1. Problem

You are faced with the requirement of performing a Q measurement at around 68 GHz. The expected Q of the device is around 2000. Your source for the measurement is a reflex klystron, which can be mechanically tuned from 68 to 73 GHz and electronically repeller-tuned over one mode. The width of a mode is about 220 MHz. You have available to you the standard laboratory equipment, such as oscilloscopes, voltmeters, etc., and microwave hardware for the 60–90-GHz range. The latter includes crystal detectors, a wavemeter, a magic tee, a slotted line, tunable shorts, waveguide, etc. No frequency standard, no microwave receiver, and no transfer oscillator are available.

The wavemeter has a loaded Q of about 800; an accuracy of $\pm$ 0.5%. Suggest a method, based on interferometric principles, for frequency calibration within a klystron mode that is sufficiently accurate for the Q measurement. If possible, the Q should be measured to an accuracy of $\pm$ 10%.

Estimate the accuracy to which your measurement is made.

21.2. Solution

It is obvious that since the Q of the wavemeter is only 800, it cannot be used to calibrate frequencies within a klystron mode with sufficient accuracy to allow measurement of Q of the order of 2000 with any degree of accuracy. For, assuming the width of the wavemeter "dip" to be about twice the width of its half-power points, then for a Q of 800 at 68 GHz, the width of the wavemeter "dip" is

$$(2)\ (68000)/(800) = 170 \text{ MHz}.$$

This is nearly equal to the width of the klystron mode.

Let us first examine to what degree of accuracy we must perform the frequency calibration. The Q to be measured is approximately 2000, with

* Problem 21 is by I. Kaufman.

an accuracy of $\pm$ 10%. Now

$$Q = f/\Delta f,$$

where f is the resonant frequency and Δf is the frequency difference to half-power points. If δf is the error in measuring f, $\delta(\Delta f)$ is the error in measuring Δf, and δQ is the error in finding Q, then

$$\delta Q = f\,\delta\left[\frac{1}{\Delta f}\right] + \left[\frac{1}{\Delta f}\right]\delta f$$

$$\delta Q = f\,\frac{-\delta(\Delta f)}{(\Delta f)^2} + \frac{\delta f}{\Delta f}.$$

This becomes

$$\frac{\delta Q}{Q} = -\frac{\delta(\Delta f)}{\Delta f} + \frac{\delta f}{f}.$$

If $|\delta Q/Q|_{\max} \leqslant 0.1$, corresponding to an accuracy of $\pm$ 10%, we must have

$$\left|\frac{\delta(\Delta f)}{\Delta f}\right| + \left|\frac{\delta f}{f}\right| \leqslant 0.1.$$

Since the wavemeter is accurate to $\pm$ 0.5%,

$$\left|\frac{\delta f}{f}\right|_{\max} = 0.005.$$

Consequently, the maximum error allowable in Δf is given by

$$\left|\frac{\delta(\Delta f)}{\Delta f}\right| \leqslant 0.1 - 0.005 \approx 0.1.$$

If the Q is to be measured to an accuracy of $\pm$ 10%, the frequency difference between half-power points must therefore also be measured to an accuracy of $\pm$ 10%. For an expected Q of 2000, we have Δf = 68,000/2000 = 34 MHz. The error in Δf allowed is therefore $\pm$ 3.4 MHz.

The interferometric method of calibration to be investigated is shown in Fig. 21.1. Here power enters the magic tee through arms 1 and 3. If the fields entering are of equal amplitude and in phase with each other, no power leaves arm S, all of the power leaves arm P. The converse is true if the fields are equal but 180° out of phase. If the frequency of the source is swept through a klystron mode, so that the relative phase of the signals of arms 1 and 3 changes through several cycles, there will be peaks and nulls in the power leaving arm S.

In the operation of the equipment, the klystron is swept through a mode (by repeller modulation), while its power output is split into two channels. One of these is used for the Q measurement, while the other is used for frequency calibration. The output patterns of the two measurements are ultimately displayed on a dual channel oscilloscope, so that the calibration derived from the frequency calibration channel is instantly available for the Q measurement (see Fig. 21.2).

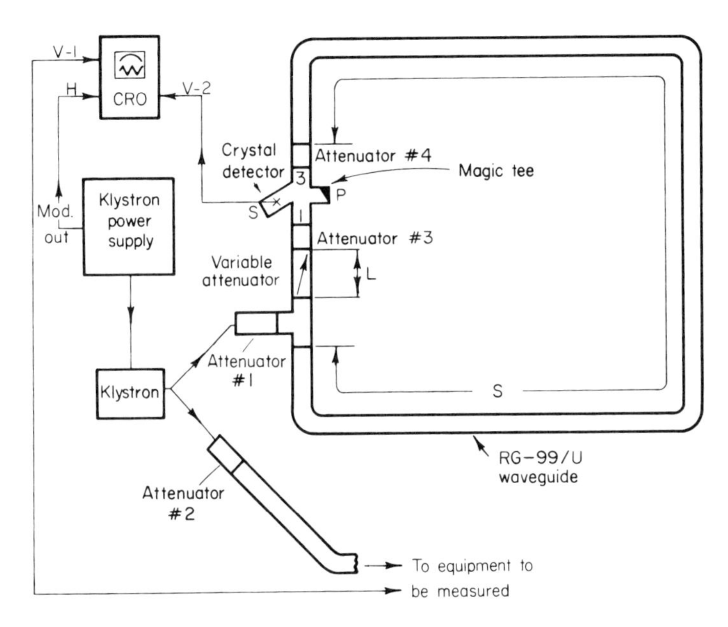

FIG. 21.1. Waveguide interferometer setup for frequency calibration, to be used in Q measurement.

In the channel used for frequency calibration, the number of guide wavelengths between arms 1 and 3 is $n = (s - L)/\lambda_g$, where λ_g is the guide wavelength. As the frequency is varied, λ_g changes, changing n. A null occurs in the power detected at S whenever n is an integer.

The change in n corresponding to a change in λ_g is

$$\Delta n = -(s - L)\frac{\Delta\lambda_g}{\lambda_g^{\,2}} = -n\frac{\Delta\lambda_g}{\lambda_g},$$

where $\Delta\lambda_g$ is the change in λ_g. For a change in the number n by one (1), $\Delta n = 1$, so that

$$\Delta\lambda_g = -\frac{\lambda_g}{n}.$$

We now relate a change $\Delta\lambda_g$ to a change in frequency $\Delta_1 f$. For single mode propagation

$$\lambda_g = \frac{\lambda_0}{(1 - (f_c/f)^2)^{1/2}},$$

where λ_0 is the free-space wavelength at frequency f and f_c is the waveguide

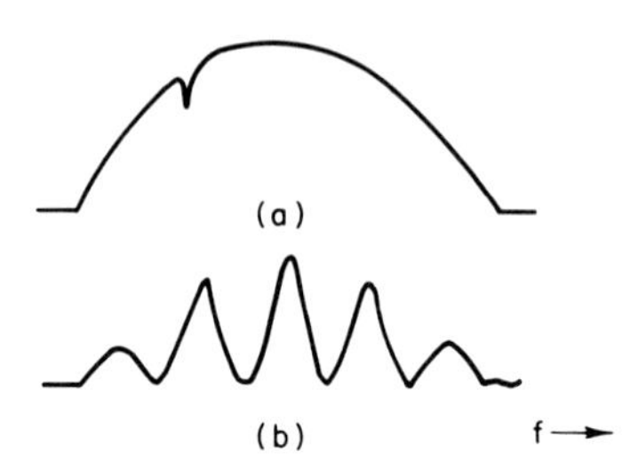

FIG. 21.2. Simultaneous display of calibration and measurement: (a) for Q measurement; (b) for calibration.

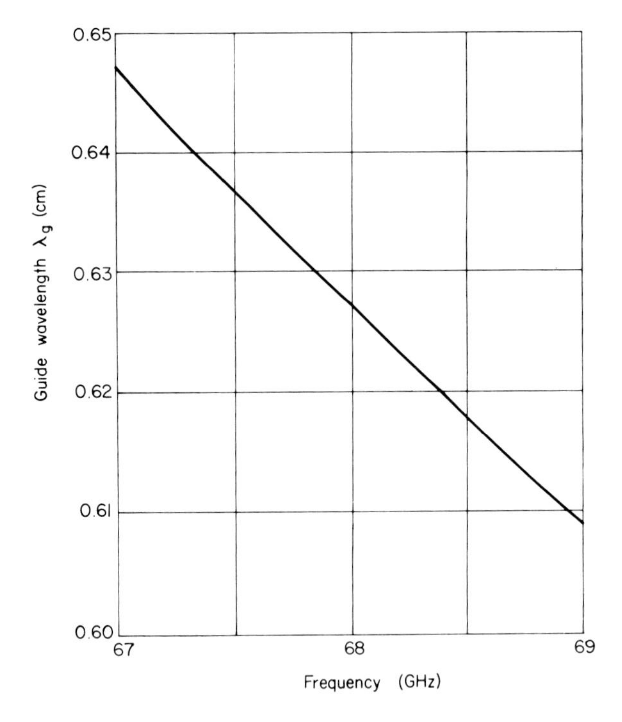

FIG. 21.3. Guide wavelength versus frequency in RG-99/U waveguide.

cutoff frequency. To arrive at the exact relation between $\Delta_1 f$ and $\Delta\lambda_g$, we can differentiate the expression for λ_g with respect to f. Even simpler, we can use a graphical method.

Suppose we used RG-99/U waveguide, whose internal dimensions are 0.310 × 0.155 cm. From the above expression, or from standard waveguide tables, the relation between λ_g and f at 68 GHz is given by Fig. 21.3. From this figure, we find, at 68 GHz, that

$$\frac{\Delta\lambda_g}{\Delta_1 f} = (1.87)(10^{-5}) \quad \text{cm/MHz}.$$

The separation between "dips" in the lower trace of Fig. 21.2 is therefore given by

$$\Delta_1 f = -\left(\frac{\lambda_g}{n}\right)\left(\frac{1}{(1.87)\,(19^{-5})}\right) = -\left(\frac{0.627}{n}\right)\left(\frac{1}{(1.87)\,(10^{-5})}\right) \quad \text{MHz}.$$

This is an expression of the n required to achieve the separation $\Delta_1 f$. As stated earlier, we require $\Delta f = 34$ MHz, with an allowable error of 3.4 MHz.

Suppose we adjust the repeller sweep and the horizontal deflection sensitivity of the oscilloscope so that a horizontal distance of 10 cm on the oscilloscope face represents 100 MHz. Since the scale on the face can be read to about 0.05 cm, our maximum error caused by the limitation of scale reading alone is then 0.50 MHz. Since this is much less than 3.4 MHz, the figure of 100 MHz/10 cm seems reasonable.

Let us require, in addition, that the separation between "dips", as in Fig. 21.2, is 40 MHz. With proper adjustment of the center frequency during the sweep, this allows three "dips" on the face so that a horizontal calibration curve can be drawn. (The peaks will help here, also.)

The separation between "dips" is $\Delta_1 f$ so that now n can be determined. From the last equation we find

$$n = (0.627)/(1.87)(10^{-5})(40.0) = 840$$

This corresponds to $(s - L) = 5.27$ m.

A waveguide loop of this length should therefore accomplish the purpose of frequency calibration.

21.3. Exercises

21.3.1.

What is the attenuation of the waveguide loop of length 5.27 m?

21.3.2.

In Fig. 21.1, the attenuators were placed to isolate the various sections from one another. The figure of 20 dB is usually considered satisfactory for isolation. In setting up the experiment with 20-dB attenuators everywhere, it will be found that the amount of power available from the usual reflex klystron (10–100 mW) is insufficient to perform the calibration in the manner described. What attenuators can be removed?

21.3.3.

What is the purpose of the variable attenuator of Fig. 21.1?

21.3.4.

The variable attenuator has an unknown amount of phase shift through it. Will this affect the accuracy of calibration seriously?

21.3.5.

Suppose the length of loop s can be measured only to within $\pm$ 0.5 cm. What effect will this have on the accuracy of the calibration?

21.3.6.

The waveguide width in the guide that you possess is accurate to within $\pm$ 25.4 μm. How does this tolerance affect the accuracy?

21.3.7.

Suppose you did not have enough length of RG-99/U waveguide, but had tapers to X-band guide (10.16 $\times$ 22.86 mm) and long lengths of this guide. Would you advise using it? Why?

21.3.8.

If instead of using waveguide you were to use a sending and a receiving antenna and a reflector on the other side of the room, how would this change the calibration?

22. MICROWAVE OPTICS*,[1]

22.1. Problem

It is desired to investigate the properties of a Fresnel zone plate lens at microwave frequencies. To be investigated are such characteristics as the ability to focus radiation, angular resolution, frequency response, and off-axis response. To this end, it has been decided to test a zone plate lens that has a transparent central region and two obstructing zones. The design frequency is to be 2.73 Gc ($\lambda = 11$ cm). The focal length is to be 10 m. Design this lens. Estimate the ratio of power density at the focus to that which would be received at this point if the lens were not present. Estimate the angular resolution. Suggest experimental methods for measuring the various quantities.

22.2. Solution

Fresnel zone plate lenses are discussed in a number of optics texts, as well as in several journal articles. We shall therefore present only a brief discussion here. For more details, the reader is advised to consult these references.[2]

The Fresnel zone plate lens operates because of Huygen's principle, i.e. that each point on a wavefront may be regarded as a new source of waves.

To design the lens, we refer to Fig. 22.1 and assume parallel radiation to enter from the left, directed perpendicular to the plane of the lens. As in other lenses, such radiation will be focused to a point on the optical axis at a distance f from the plane, where f is the focal length. This focal point is point F in Fig. 22.1. If the radiation is to be focused at F, then the wavelets that are radiated from the secondary sources in the plane of the lens (plane L-L') must all add in phase at F.

[1] This problem was suggested by the work of H. G. Oltman, Jr., in A Fresnel half-period zone plate for focusing electromagnetic energy in the one meter wavelength region, Thesis. The University of New Mexico, 1954.

[2] A lucid introductory discussion is given by F. A. Jenkins and H. E. White, in 'Fundamentals of Optics', Third Edition, McGraw-Hill, New York. 1957.

* Problem 22 is by I. Kaufman.

The phase angle of travel from plane L-L' to F is given by

$$\phi = \frac{\omega}{c} D = \frac{2\pi}{\lambda} D,$$

where ω is the angular frequency, c is the velocity of light, D is the distance,

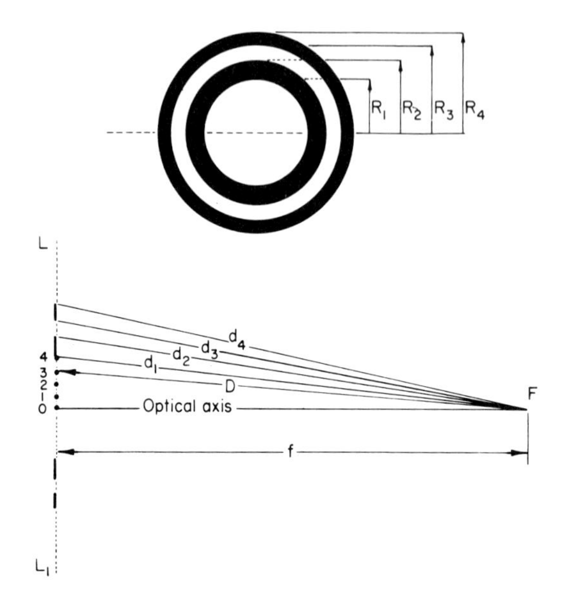

FIG. 22.1. Fresnel zone plate lens with two blocked zones. Top: front view; bottom: side view, showing distances to focal point.

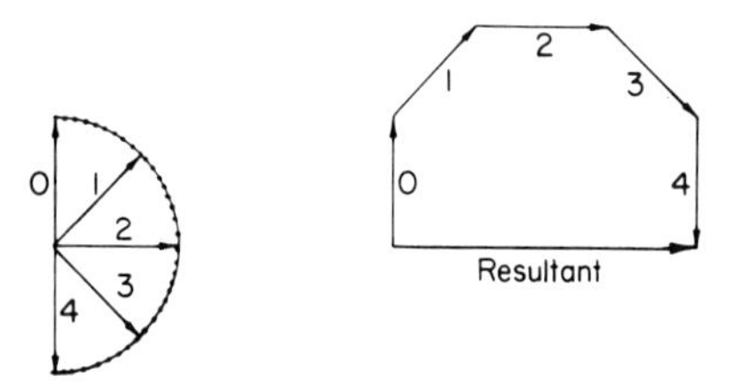

FIG. 22.2. Phase relations of field components radiated from points 0, 1, 2, 3, and 4 of Fig. 22.1 (left) and addition to form resultant field (right).

and λ is the wavelength. The phase angle for the central ray is therefore $\phi = (2\pi/\lambda)f$. While a true lens is shaped so that the radiation from every secondary source is in phase with every other, this situation can obviously not be true for the zone plate. Instead, the phase relations are as in Fig. 22.2. Here the radiation received at F from several secondary sources (points 0, 1, 2, 3, and 4) within the first zone is plotted in terms of the phase relations of the (electric) field received at F from these sources. It is seen that the radiated fields are definitely not in phase, but they do have in-phase components. The edge of a zone is therefore at the radius at which these components reverse in sign. Since the radiation from this radius outward to another radius stays reversed in sign, the zone is blocked by the designer from there until the radius at which the radiation adds to the radiation from the central zone again.

In this manner, the distances d_i which define the zones, as in Fig. 22.1, are given by

$$\frac{2\pi}{\lambda}(d_i - d_{i-1}) = \pi.$$

We therefore have here

$$\frac{2\pi}{\lambda}(d_1 - f) = \pi;$$

$$\frac{2\pi}{\lambda}(d_2 - d_1) = \pi;$$

$$\frac{2\pi}{\lambda}(d_3 - d_2) = \pi;$$

$$\frac{2\pi}{\lambda}(d_4 - d_3) = \pi.$$

But

$$d_1 = [f^2 + R_1{}^2]^{1/2}.$$

Substituting $\lambda = (11.0)\ 10^{-2}$ m, $f = 10.00$ m, we find $R_1 = 1.05$ m. Continuing, we obtain the values of Table I.

TABLE I.
Radii for Zone Plate of Two Blocked Zones with $\lambda = (11.0)\ 10^{-2}$m; $f = 10.00$ m.

Radius	(m)
R_1	1.05 m
R_2	1.49 m
R_3	1.82 m
R_4	2.11 m

The values of Table I are all the data required for the design of the lens.

To find the ratio of power density at the focus to that which would be received at this point if the lens were not present, we must make a calculation of the power radiated from the Huygens sources that exist at plane L-L' when a plane wave strikes this plane from the left. (If no zones were blocked, this would, of course, result in the intensity of the plane wave again.)

In a simplification, in which we do not actually carry out the integration required by the complete analytical solution, we can consider that the intensity of the electric field at F that is due to the radiation from each zone is given by

$$E_m = C \frac{A_m}{s_m} (1 + \cos \theta_m),$$

where A_m is the area of the zone, s_m is the average distance of the zone from F, C is a constant, and θ_m is the angle between the line of length s_m and the optical axis.

From this approximation and a summation of the contributions from all zones that would exist if plane L-L' were divided into an infinite number of zones (none blocked), it can be shown that the contribution from the central zone, E_1, is $E_1 = 2E_0$, where E_0 is the field intensity of the incoming wave. (See Jenkins and White, *op. cit.*)

Returning to our lens, we note by numerical substitution in the intensity equation above that the contributions from each of the four zones (two open and two blocked) of the lens are approximately equal. If we assume, additionally, that only half of the next zone beyond the outermost blocked zone contributes, we find

$$E \simeq 2E_0 + 2E_0 + E_0 = 5E_0.$$

The electric field at the focus should therefore be about five times that which would exist if there were no lens. The power density is therefore increased 25 times.

To estimate the angular resolution, we assume that the usual aperture formula for Fraunhofer diffraction holds. This means that the Fresnel zone plate lens can resolve two sources if they are separated in angle at the lens by more than

$$\alpha = 1.22\lambda/K,$$

where α is the angle, in radians, λ is the wavelength, and K is the lens diameter.

Since we do not exactly know what diameter to use here, we will merely set an upper limit by choosing as our lens diameter the inner diameter of the outer blocked zone. The angular resolution expected is therefore

$$\alpha \simeq (1.22)(11)(10^{-2})/(2)(1.82) = 0.037 \text{ rad} = 2.1°.$$

The construction of the lens is quite simple. Since the lens is composed merely of the two concentric rings that block the radiation, it is only necessary to find some means of mounting these rings.

Since the purpose of the rings is to block the radiation, and not necessarily to absorb it, it will not be necessary to use microwave absorbing material. Instead, sheet metal or wire screen will suffice.

A good method of mounting these rings is on a framework of wood, or some other nonconductive material. In a successful model, constructed by H. G. Oltman, Jr., the rings were secured to a wooden frame by heavy construction tape, which had fiberglass imbedded in it.

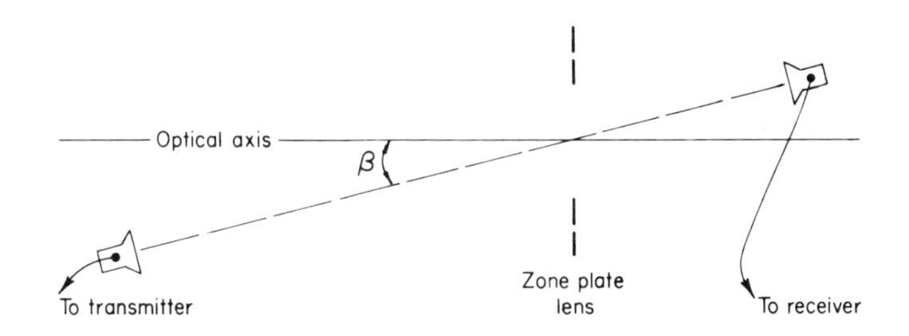

FIG. 22.3. Experimental arrangement for testing Fresnel zone plate lens.

To test the lens, it would seem desirable to test it with a source of parallel 11-cm radiation. In practice, this may not be possible. Recalling that the device is a lens, however, suggests that a point source of radiation at some distance from the lens could be used just as well. The image formed would then not be at the focal distance, but at a location given by the thin lens formula. By placing a receiving antenna at this point, connected to a receiver, the effective concentration power of the lens can be measured. The diagram of Fig. 22.3 is appropriate here.

Other quantities to be measured are the angular resolution and the useful field of the lens. By swinging the receiving antenna in an arc about the center of the lens, at the distance of the image, the angular resolution is obtained.

We can define the useful field of the lens as twice the included angle between the optical axis and the direction to the source of the radiation for which the maximum power received by the receiving antenna is within half of the maximum power received when the source is on the optical axis. Consequently, by moving the direction of the source (angle β in Fig. 22.3), this field can be measured.

22.3. Exercises

22.3.1.

In the construction of the wooden frame that holds the obstructing portions of the lens it is found necessary to use nails. In what direction should these point in order to interfere least with the lens action?

22.3.2.

In the test of the performance, shown in Fig. 22.3, is there any restriction to the beam width of the transmitting antenna? Is there any restriction to the distance between this antenna and the lens?

22.3.3.

In the test of performance, what restriction is there on the beam width of the receiving antenna?

22.3.4.

Suppose this zone plate lens is to be used with a small antenna at its focus to transmit radiation. What beam width would you choose for this small antenna to make maximum use of the lens?

22.3.5.

Design a Fresnel zone plate lens, whose innermost zone is obstructed and which has three additional obstructed zones. Compute the parameters that were calculated for the lens in the problem, above, for your lens.

22.3.6.

We have found that the power density of our lens, above, should be about 25 times that which would be received if the lens were not present. Suppose that instead of the zone plate, we had designed a perfect lens (or parabolic reflector), which has the same diameter as the diameter of the outer obstructed zone of our zone plate. The radiation focused by this lens is all absorbed by a small antenna, just as in our case. How much better would such a perfect lens be? Why?

22.3.7.

In the experimental test of our lens, the transmitting antenna is located 35 m from the zone plate. Where would the energy from this source be focused? What is the minimum distance between transmitting antenna and zone plate that is allowed so that good focusing will still occur?

22.3.8.

What happens to the focal length of our lens as the frequency is changed? If we define the bandwidth of the lens as that band over which the power density of the lens at the midband focus drops to one-half of its maximum value, what is the approximate bandwidth of our lens?

22.3.9.

The focusing ability of a zone plate lens can be increased by utilizing the areas that were obstructed in filling them with dielectric material. How thick would a dielectric material of dielectric constant $\varepsilon_r = 4$ and relative permeability $\mu_r = 1$ have to be, to achieve maximum improvement? How much would the ratio of power density at the focus to that without lens improve? Is it possible to improve the focusing any more by zone subdivision?

23. MEASUREMENT OF GRAVITATIONAL RED SHIFT*

23.1. Introduction

One of the classical tests for the general theory of relativity is based upon the observation of gravitational red shifts of spectral lines produced on stars or on the sun. Because of pressure broadening effects occurring in lower (dense) layers of the stellar atmosphere, it would appear advantageous to select lines formed in outer layers, i.e., strong lines in which the absorption coefficient is large. Furthermore, the lines should lie in the visible range of the spectrum to be observed from the ground with good sensitivity. Hydrogen lines and, to a lesser extent, helium lines turn out to be not suitable because of their large widths stemming from thermal Doppler and intermolecular Stark effects. Of the other strong lines, the sodium D lines appear to be good candidates, being both strong and reasonably sharp.

23.2. Problem

Selecting thus, say, the strongest component of the sodium doublet for observation, the task is to design apparatus capable of yielding a measurement of the solar gravitational red shift to, e.g., 10% accuracy.

23.3. Solution

23.3.1. Preliminaries

As a first step, the expected effect should be estimated from the following data: solar mass $\approx 10^{33}$g, radius $\approx 10^{11}$cm, gravitational constant $\approx 10^{-7}$ dyn cm^2/g^2. Other preparatory estimates are those of Doppler, pressure, and natural line broadening (atom density $\approx 10^{16}$cm^{-3}, temperature $\approx$ 6000°K) and of possible competing shift mechanisms due, e.g., to the solar rotation (27 day period) or the above line broadening mechanisms.

23.3.1.1. Results. Expected effect $\approx$ 10 mÅ; thermal Doppler width $\approx$ 40 mÅ; pressure broadening $\approx$ 0.1 mÅ (negligible); natural width $\approx$ 0.1 mÅ (negligible); solar rotation shift $\lesssim$ 50 mÅ, but pressure shifts, etc., will be negligible.

* Problem 23 is by Hans R. Griem.

23.3.1.2. Conclusions. Observations must be made as to avoid seeing the equatorial regions of the solar limb, and measurements of the shift to within $\sim 1\%$ of the expected line width which will actually be larger than estimated here over most of the disk because of self-absorption. This suggests therefore to observe the polar caps (or to make measurements at points symmetrical to the axis to balance out the rotational effect).

23.3.2. Telescope

Describe requirements in regard to angular resolution, tracking, and optical speed.

23.3.2.1. Result. To keep solar rotational shifts to $\lesssim 1$ mÅ, angular resolution should be $\sim 1\%$ of the angular size of the sun, i.e., $\sim 10^{-4}$ rad, and a $\sim 10^4 \times \lambda \approx 1$ cm objective would suffice in principle. However, for reasons of good signal a considerably larger objective is desirable, with a focal length of ~ 1 m as to produce a primary solar image of ~ 1 cm diameter. The focal plane would then contain an opaque screen with an ~ 0.1 mm pinhole admitting light from one of the polar caps to the spectrograph or interferometer. Tracking to the 10^{-4} rad accuracy will not be easy, but not impossible either.

23.3.3. Spectrograph or Interferometer

Estimate the resolving power necessary and how it might be realized by a grating or anetalon instrument.

23.3.3.1. Result. Shifts can be measured to within $\sim 10\%$ of apparatus width, need therefore $\sim 10^6$ resolving power and 0.1 Å/mm reciprocal dispersion (assuming one-to-one optics in spectrograph. Echelle grating . . . Fabry-Perot interferometer . . .).

23.3.4. Detectors

Would photographic or photoelectric detection be desirable, and what is the exposure or scanning time expected for your instrument?

23.3.5. Wavelength Calibration

How would you calibrate your instrument?

23.3.5.1. Answer. Isotope lamp with microwave excitation imaged essentially through the same optical system.

24. RELATIVISTIC EFFECT ON ATOMIC CLOCKS*

24.1. Problem

Atomic clocks (cesium frequency standards) can now be made small enough to become "portable." That means that for purposes of intercomparison one cesium standard can be transported over long distances from one laboratory to another and brought back to its point of departure for examination for possible drift or other effects. Assuming that a clock was sent by air at an altitude of 10^4 m, with a ground speed of 10^6 m/h, for a total time of 100 h; estimate the relativistic effects on the moving clock with respect to the stationary clock. Is it measurable under the given experimental conditions if the random errors and drifts in the clock rates are about one part in 10^{12}?

24.2. Solution

The effects must be considered from the viewpoint of general relativity, since relative motion, rotation, and gravitation are all involved. A flying clock moving at constant velocity v and constant altitude h will indicate an elapsed time T_e, different from that of a stationary clock, T_0, on the ground by a fractional amount:

$$\frac{T_e - T_0}{T_0} \approx \frac{gh}{c^2} - \frac{v^2}{2c^2}$$

where g is the acceleration of gravity (including the effect of rotation of the earth) and c is the velocity of light. In this approximation the gravitational and motion effects may be separated. The first term is the gravitational shift, which is "blue" since the flying clock is farther from the earth than the stationary clock. The second term is the "time dilatation" of special relativity caused by the relative velocity.

* Problem 24 is by L. Marton.

Taking the above numbers:

$$\frac{T_e - T_0}{T_0} \approx \frac{9.81 \times 10^4}{(3 \times 10^8)^2} - \frac{(2.77 \times 10^2)^2}{2(3 \times 10^8)^2}$$

$$= 1.09 \times 10^{-12} - 0.43 \times 10^{-12} = 0.66 \times 10^{-12}.$$

Thus the "blue shift" term is 1.09×10^{-12} and the time dilatation term is -0.43×10^{-12}, or a total of 0.66×10^{-12}. The two terms give an elapsed time difference for 100h

$$T_e = 3.6 \times 10^5 + 2.4 \times 10^{-7} \quad \text{sec.}$$

We see that, although the "blue shift" term is a trifle above the limit given for random errors and drifts in the clock rates, as the "time dilatation" term has to be subtracted, the total expected effect is below what is observable.

24.3. Exercises

1. Demonstrate the limits of validity of the above approximate equation.
2. Calculate the lower limits of heights and velocities beyond which relativistic effects can not be neglected. To do this it is useful to read S. F. Singer's article: Application of an artificial satellite to the measurement of the general relativistic "red shift", Phys. Rev. **104,** 11, 1956. His final equation, which applies here, can be rewritten as:

$$\frac{T_1 - T_2}{T_1} = \frac{gM}{2c^2 r}\left(\frac{1 - 2h/r}{1 + h/r} - \frac{3\omega^2}{4\pi g \rho_M}\right).$$

M is the mass of the earth, ρ_M is the average density of the earth, r is its radius, h is the height assumed for a circular orbit above the earth surface, and ω is the angular velocity of the earth. (In comparing with Singer's equation, watch for proper use of dimensions.)

SUGGESTED REFERENCES

"The Feynman Lectures," Vol. I. Addison-Wesley, Reading, Massachusetts, 1963.
Møller, C., "The Theory of Relativity." Oxford Univ. Press, London and New York, 1952.
Pauli, W., "Theory of Relativity." Pergamon Press, Oxford Univ. Press, London and New York, 1958.

25. ELECTRON STORAGE RING*

25.1. Problem

An electron storage ring is a device where high energy electrons (or positrons) can be kept by a suitable magnetic field on a circular orbit for times exceeding a full day. The primary requirement is freedon from losses through scattering processes, thus ultra high vacuum is one necessary condition. There are, however, energy losses due to radiation by the orbiting electron. Calculate the magnitude of such losses, together with such characteristics of the emitted electromagnetic radiation, which may make it a useful tool for physical research.[1-8]

25.2. Solution

A relativistic electron, moving on a circular orbit of radius ρ, emits light, highly polarized in the plane of the orbit. The total power radiated by one electron is equal to:

$$P_\gamma = \frac{2}{3}\frac{r_e mc^3\gamma^4}{\rho^2} \tag{25.1}$$

where m is the rest mass of the electron, r_e is its classical radius ($r_e = e^2/mc^2 = 2.818 \times 10^{-13}$ cm),

$$\gamma = \frac{1}{(1-\beta^2)^{1/2}} \quad \text{with} \quad \beta = v/c,$$

[1] D. H. Tomboulian, and P. L. Hartman, *Phys. Rev.* **102,** 1423 (1956).
[2] D. H. Tomboulian, and D. E. Bedo, *J. Appl. Phys.* **29,** 804 (1958).
[3] K. Codling, and R. P. Madden, *J. Appl. Phys.* **36,** 380 (1965).
[4] R. P. Madden, D. L. Ederer, and K. Codling, *Appl. Optics* **6,** 31 (1967).
[5] J. A. R. Samson, "Techniques of Vacuum Ultraviolet Spectroscopy." Wiley, New York. 1967.
[6] A. Blanc-Lapierre, Onde Electrique, no. 471, June 1966.
[7] H. Bruck, "Accelerateurs Circulaires de Particules." Presses Universitaires de France, Paris, 1966.
[8] J. Schwinger, *Phys. Rev.* **75**, 1912 (1949).

* Problem 25 is by L. Marton and A. Blanc–Lapierre.

v is the speed of the electron, c is the speed of light. The energy radiated per revolution is:

$$\Delta E_\gamma(\text{keV}) = 8.85 \times 10^2 \frac{E^4(\text{GeV})}{\rho(\text{cm})}. \tag{25.2}$$

For instance, for an energy E of 450 MeV and a radius ρ of 110 cm ($B =$ 13,500 G), ΔE is about 3.3 keV.

Synchrotron radiation is extremely pointed in the forward direction. With good approximation it can be said that the radiation is emitted essentially in a cone, whose axis coincides with the direction of the moving electron, its half angle being equal to:

$$\delta = \frac{1}{\gamma} = \frac{m_0 c^2}{E}. \tag{25.3}$$

As an example, for $E \approx 0.5\text{GeV}$, $\delta \approx 10^{-3}$ rad.

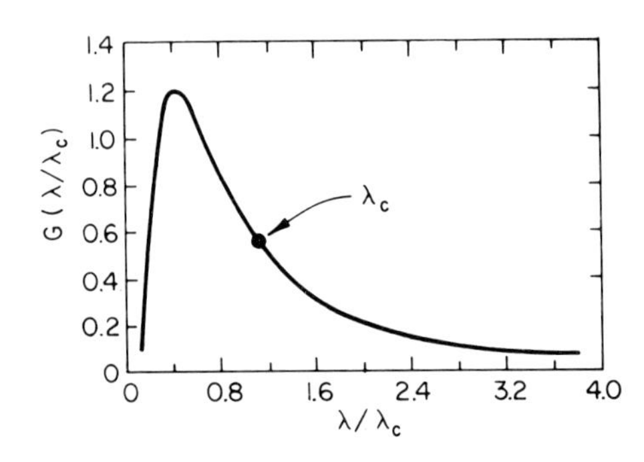

FIG. 25.1. Universal spectral distribution curve for the radiation from monoenergetic electrons.

The spectral distribution of the synchrotron radiation depends upon the electron energy and orbital radius. A simple universal expression for the power radiated per angstrom at wavelength λ into all angles, for a single monoenergetic electron of energy E traveling an orbit of radius ρ, is:

$$\bar{P}(\lambda) = \frac{3^{5/2}}{8\pi^3} \frac{e^2 c}{\rho^3} \left(\frac{E}{m_0 c^2}\right)^7 G\left(\frac{\lambda}{\lambda_c}\right) \tag{25.7}$$

where $G(\lambda/\lambda_c)$ is the function plotted in Fig. 25.1, with λ_c given by

$$\lambda_c = \frac{4\pi\rho}{3} \left(\frac{m_0 c^2}{E}\right)^3$$

or

$$\lambda_c(\text{Å}) = 0.559 \frac{\rho\ (\text{cm})}{E^3(\text{GeV})}.$$

Above the "critical" wavelength λ_c the radiation decreases very rapidly. Using our earlier example, i.e. $\rho = 110$ cm, $E = 0.45$ GeV, λ_c is about 67.5 Å.

Figure 25.2 shows the spectral distribution for a few characteristic examples, such as 0.24, 0.26, 0.28, 0.30, and 0.32 GeV, for a radius of approximately 1 m.

Synchrotron radiation is emitted in a discontinuous fashion of individual photons. Each act of emission produces a small change in the particle energy and an extremely minute change in its direction.

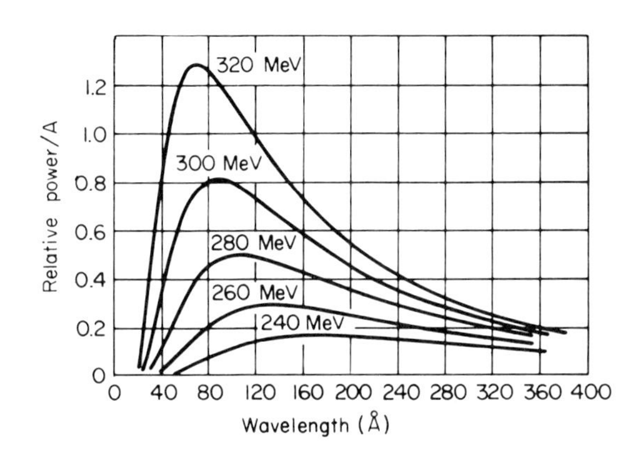

FIG. 25.2. Comparison of the relative spectral distributions for the radiation from monoenergetic electrons at various energies. This illustrates the sensitive variation of the distribution with energy.

The light emitted in the direction of the motion of the electron produces on this latter, by reaction, a force in the opposite direction. As a consequence of the synchrotron radiation storage rings have the following characteristics:

1. the energy is very precisely defined;
2. the beam transversal dimensions are small;
3. the angular divergence of the beam is very small.

The following estimates illustrate these last three items. The energy fluctuations can be identified with the energy loss due to synchrotron radiation during a time equal to the time required for the damping of oscillations. For our earlier example of $E = 0.45$ GeV, $B = 13{,}000$ G, $\rho = 110$ cm, the calculated damping constant is $\tau = 0.02$ sec. The energy loss per revolution is, as we have seen, $\Delta E = 3.3$keV. The frequency of revolution being 13.62 MHz, the energy lost during the damping time is $3300 \times 13.62 \times 10^6 \times 0.02 \approx 9 \times 10^8$ eV. By assuming that the average energy carried by the photon is of the order of 100 eV, the number of photons emitted during τ sec becomes $\bar{n} = 9 \times 10^6$. These photons are *spread in time* following a Poisson distribution; thus $\overline{\Delta n^2} = \overline{(n - \bar{n})^2} = \bar{n}$. The energy loss fluctuation during τ sec becomes thus $\sqrt{\bar{n}}\, h\nu$ and this fluctuation alone accounts for the evaluation of the beam energy fluctuation ΔE as the average loss $\bar{n}h\nu$ is compensated by the gain from the high frequency. Therefore

$$\Delta E \approx \sqrt{\bar{n}}h\nu \approx 3 \times 10^3 \times 100 = 3 \times 10^5 \text{ eV};$$

that is, the energy spread is of the order of one part in 1500.

It is interesting to calculate the order of magnitude of the visible radiation perceived by the human eye. Let us assume the mean energy of 150 MeV for the electrons and a radius of curvature of 100 cm. The energy loss per revolution is about 40 eV [Eq. (25.2)]. Let us assume also that the time required for one turn is 7×10^{-8} sec. The total current, corresponding to *one* circulating electron is equal to $(1.6 \times 10^{-19})/(7 \times 10^{-8})$ A. Thus the radiated power per electron is of the order of 10^{-10} W. The eye sees only a part of this, as the electron is only a short time in a portion of the orbit, where the narrow cone of emission is in line with the pupil of the eye. In assuming the solid angle subtended by the eye as being 3×10^{-3} rad and assuming also that 10% of the total emissive power is in the visible region, we come to the conclusion that from a *single* electron the eye receives a total power of about 10^{-15} W.

26. MASS SPECTROMETER FOR UPPER ATMOSPHERE RESEARCH*

26.1. Problem

For use in upper atmospheric research a magnet-free mass spectrometer, with a resolution of approximately $\Delta M/M = 1/200$, is needed. Employ an electrostatic type of instrument with high frequency modulation. Discuss the status of magnet-free instruments from the literature and give design principles for a parabolic-path-type instrument.

26.2. Solution

26.2.1. Introduction

At present, several types of nonmagnetic time-of-flight mass spectrometers are in operation. Examine the literature to determine the principles of operation, operating conditions, and mass resolutions for the following instruments: the Bendix-type instrument[1,2], the Bennet-type instrument [3,4] the Farvitron tube[5], and the quadrupole mass filter[6-8].

26.2.2. Principle of the Electrostatic Time-of-Flight Mass Spectrometer[9]

The electrostatic spectrometer, like the Bennet-type instrument, employs a set of rf fields and drift spaces. The main difference is that by means of parabolic ion paths, high energy resolution is achieved (Fig. 26.1). A continuous parallel ion beam of constant energy first passes a "modulator" $M1$, consisting of three parallel electrodes. A proper rf voltage is applied to the middle electrode of the modulator and the outer ones are connected

[1] A. E. Cameron, and D. F. Eggers, Jr., *Rev. Sci. Instr.* **19,** 605 (1948).

[2] W. C. Wiley, and I. H. McLaren, *Rev. Sci. Instr.* **26,** 1150 (1955).

[3] W. H. Bennet, *J. Appl. Phys.* **21,** 143 (1950).

[4] K. R. Ryan, and J. H. Green, The performance of a radio frequency mass-spectrometer of the bennet type at elevated pressures. Dept. of Nuclear and Radiation Chemistry, Univ. of New South Wales, Kensington, Australia.

[5] Z. Tretner, *Angew. Phys.* **11,** 395 (1959); **14,** 23 (1962).

[6] Paul and Steinwedel, *Z. f. Naturf.* **8a,** 448 (1953).

[7] U. von Zahn, *Z. Physik* **168,** 129 (1962).

[8] Brunnee-Voshage "Massenspektrometrie." Verlag K. Thiemig KG, Munich, 1964.

* Problem 26 is by M. J. Higatsberger.

to ground. Depending on the phase of the rf voltage, part of the ion beam is accelerated or decelerated. Then the ion beam enters a homogeneous electrostatic field at an entrance angle α. Show that the ions follow parabolic paths with a range depending on the kinetic energies of the ions. Only ions with maximum energy gain at $M1$ can pass slit $S2$. After this slit a modulator $M2$ is mounted to which the same rf voltage is applied. The arriving ions again are phase-dependent accelerated. Once more, only those which have obtained the maximum possible energy in $M2$ are able to pass the slit $S3$ in modulator $M3$, to which the same rf voltage is fed. After having

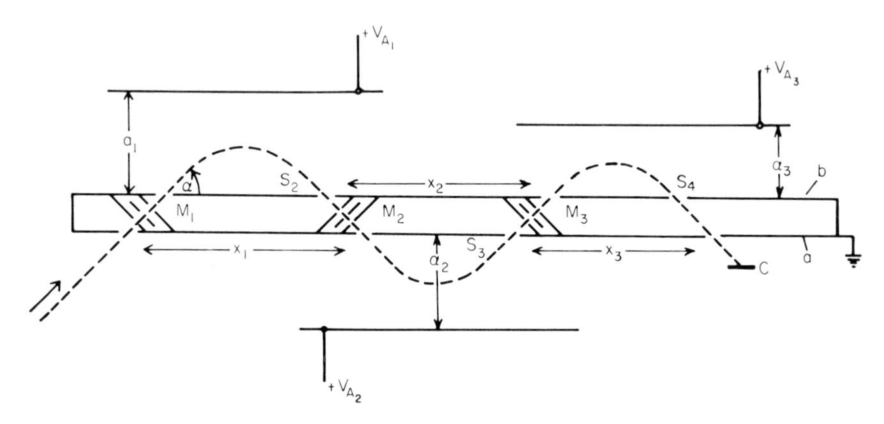

FIG. 26.1. Electrostatic time-of-flight spectrometer.

passed the last energy selecting slit $S4$, the ions are recorded by the collector C. Show that the time-of-flight along the parabolas is different for particles of different masses. Thus mass discrimination is achieved by varying the rf frequency. To improve the resolution a greater number of modulators and drift spaces may be employed.

26.2.3. Calculations

26.2.3.1. Phase Condition and Energy Gain in a Single Modulator—Bennet[3] and Smith and Damm[10] have discussed the phase condition in the modulator for the maximum energy gain. Let us consider an ion of mass m and initial energy per unit charge V_B (linear velocity v_0) which passes the

[9] M. J. Higatsberger, "An Electrostatic Time-of-Flight Mass Spectrometer," p. 237. Nuclear Masses and Their Determination, (H. Hintenberger, ed.), Pergamon Press, Oxford Univ. Press, London and New York, M. J. Higatsberger *et al.*, "Advances in Mass Spectrometry," Vol. 3, p. 803, The Institute of Petroleum, London, 1966.

[10] L. G. Smith and C. C. Damm, BNL 2740, 1956.

modulator [Fig. 26.2(a)]. The rf voltage $V = V_0 \sin(\omega t + \Phi)$ is applied to the central electrode. After leaving the modulator, the ion has the energy per unit charge $V = V_B + \Delta V$. The terms d_1 and d_2 are the distances between the electrodes and s and s' are the slit widths. For the following calculation it is assumed that:

$$\begin{aligned} &\text{(i)} \quad V_0 \ll V_B \\ &\text{(ii)} \quad b \ll d_1, d_2 \\ &\text{(iii)} \quad s, s' < d_1, d_2. \end{aligned} \tag{26.1}$$

Φ is the phase angle of the rf voltage and ϕ is the phase shift of the ion with

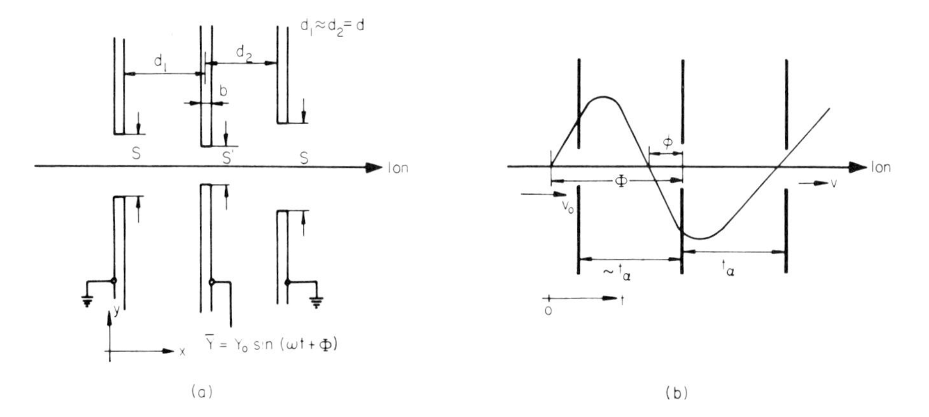

FIG. 26.2. The single modulator.

respect to the rf voltage on the median electrode [Fig. 26.2(b)]. They are connected by the relation

$$\Phi = \pi - \phi. \tag{26.2}$$

Show that the energy gain for a charge e in a modulator is given by:

$$\Delta W = e\,\Delta V \approx v_0\,\Delta(mv) = \frac{2eV_0 v_0}{\omega d} \cos\phi \left(1 - \cos\omega \frac{d}{v_0}\right) \tag{26.3}$$

or

$$\Delta V = \frac{2V_0 v_0}{\omega d} \cos(\omega\,\Delta t)\left[1 - \cos\omega \frac{d}{v_0}\right]$$

where Δt is the time difference between rf phase $\Phi = 0$ and the time of the particle transit at the medium electrode. Hence, ΔV has a maximum for

$\Delta t = 0$. For ions of different masses, show that Eq. (26.3) can be written as:

$$\Delta V = \frac{V_0 v_0}{\omega d} [\cos \Phi - 2 \cos (\omega t + \Phi) + \cos (2\omega t + \Phi)]. \quad (26.4)$$

Show that Eq. (26.4) is a maximum with respect to variations in Φ for

$$\omega t + \Phi = \pi. \quad (26.5)$$

What is the physical significance of this result? Show that Eq. (26.4) is a maximum with respect to variations in ω for

$$(\pi - \Phi) \sin \Phi - \cos \Phi - 1 = 0. \quad (26.6)$$

Evaluate Eq. (26.6) numerically for $\Phi = \Phi_m$. With this value show that the maximum possible energy gain is $\Delta V_m = 1.4492\ V_0$. A combination of Eq. (26.5) with the optimum value for Φ gives:

$$\omega \frac{d}{v_0} = \omega t = \pi - \Phi_m = \psi. \quad (26.7)$$

Show that this equation may be written in the form

$$\nu = \frac{k}{\sqrt{M}} \quad (26.8)$$

where M is the mass of the ion. Evaluate the quantity k. Equation (26.8) gives the relation between mass and frequency, hence the instrument has a quadratic mass scale. Show that ions of different mass under optimal conditions gain the energy (for ω constant)

$$\Delta V = \frac{4V_0}{\psi} \sqrt{\frac{M_0}{M}} \sin^2 \left(\frac{\psi}{2} \sqrt{\frac{M}{M_0}} \right) \quad (26.9)$$

where M_0 is the mass with maximum energy gain. Plot $\Delta V/V$ as a function of M/M_0 for $0.01 \leqslant (M/M_0) \leqslant 100$. (Use log scale for M/M_0.) Comment on the mass discrimination for a single modulation.

26.2.3.2. Ion Paths in the Drift Spaces. Show that positive ions with an energy $[eV_B]$ which enter a homogeneous electrostatic field at the angle α (see Fig. 26.1) describe parabolas with the range:

$$x = \frac{2aV_B}{V_A} \sin 2\alpha \quad (26.10)$$

where V_A and a are the deflecting potential and distance, respectively between the deflecting and grounded electrodes. If a slit S with a width

$s = \delta x$ is provided in the grounded electrode, only ions within the energy interval

$$\delta V < \frac{\delta x}{x} V \tag{26.11}$$

may pass.

Show that the time of flight along the parabolic path depends on the mass according to

$$\tau = f\,(MV_B)^{1/2}. \tag{26.12}$$

Evaluate the quantity f. One notes from Eq. (26.11) that any type of directional focusing would result in differences of the flight time, which should be avoided in this apparatus. Calculate variation of the flight time from Eq. (26.11) and show that the phase shift ϕ in Eq. (26.3) is given by

$$\phi = \omega\,\delta\tau = \frac{\omega\tau}{2}\left(\frac{\delta M}{M} + \frac{\delta V_B}{V_B}\right). \tag{26.13}$$

26.2.3.3. Phase Condition and Energy Gain in a Multimodulator System. As can be seen from Eq. (26.3) only those ions can gain maximum energy ΔV_m which pass the medium electrode of a modulator at phase angle $\psi = 0$. In order to satisfy this condition it is necessary that the time of flight between two modulators is an integer multiple n of the rf period

$$\tau = nT \qquad (n = 1, 2, 3, \ldots). \tag{26.14}$$

Show that the width of the parabolic paths becomes

$$x_i = \frac{2\pi n_i\, d_{i+1} \cos\alpha}{\psi}, \tag{26.15}$$

and the potentials in the deflecting electrodes may be expressed by

$$V_{Ai} = \frac{2\psi a_i V_B \sin\alpha}{\pi n_i\, d_{i+1}}, \tag{26.16}$$

where the index i marks the number of the modulator.

For a spectrometer with k modulators the frequency is adjusted in such a way that for ions with mass M Eq. (26.8) holds. In the first modulator the ions gain additional energy given by Eq. (26.9). All masses deviating from M_0 gain less than the maximum energy. Hence, a variation in the time of flight between the first and the second modulator arises [see Eq. (26.13)]. Therefore, show that the energy gain in the jth modulator is:

$$\Delta V_j(M) = \frac{4V_0}{\psi}\left(\frac{M_0}{M}\right)^{1/2}\left(\frac{V_B\sum_1^{(j-1)}\Delta V_i(M)}{V_B + (j-1)\,\Delta V_m}\right)^{1/2}\cos\left[\sum n_i\pi\left(\frac{M - M_0}{M_0}\right.\right.$$

$$\left.\left. + \frac{(j-1)\Delta V_m - \sum_1^{(j-1)}\Delta V_i(M)}{V_B + (j-1)\,\Delta V_m}\right)\right]\sin^2\frac{\psi}{2}\left(\frac{M}{M_0}\right)^{1/2}$$

$$\times\left(\frac{V_B + (j-1)\,\Delta V_m}{V_B + \sum_1^{(j-1)}\Delta V_i(M)}\right)^{1/2}. \tag{26.17}$$

After the last modulator the ions describe a parabola and have to pass the exit slit S_k with the width s_k. Since there is no time condition, the length of the last parabola is arbitrary. According to Eq. (26.11) only ions within the energy interval

$$\delta V = k\,\Delta V_m - \sum_1^k \Delta V_i(M) \leqslant \frac{s_k}{x_k}(V_B + k\,\Delta V_m) \tag{26.18}$$

can reach the collector.

In order to fulfill Eq. (26.6) for each modulator with the same value of ω, the distance d between the modulator electrodes has to be adapted for the increasing velocity v_0 according to

$$\frac{d_1}{d_i} = \frac{v_{1c}}{v_{ic}} = \left(\frac{V_B}{V_B + [(i-1)/2]\,\Delta V_m}\right)^{1/2}. \tag{26.19}$$

This correction improves the resolution and already is taken into account in Eq. (26.17). Another correction is made necessary by the continuous operation of the ion source which causes an additional phase shift ψ, where

$$|\,\phi\,| = \text{arc cos}\left(\frac{j\,\Delta V_m - (V_B + j\,\Delta V_m)\,(s_j/x_j)}{\sum \Delta V_j\,(M)}\right). \tag{26.20}$$

In Eq. (26.17) the term $\pm\,|\,\phi\,|$ has to be added to the argument of the cosine function.

26.2.3.4. Resolution. A rough calculation with Eq. (26.17) shows that for $M \approx M_0$ only the cosine term influences the variation of $\Delta V_j\,(M)$ markedly. Using Eq. (26.13), (26.17), and (26.19) show that

$$\psi + \bar{\phi} = \pi\sum n_i\left(\frac{\delta M}{M} + \frac{\delta V}{V}\right). \tag{26.21}$$

With $\delta M/M \ll 1$ and $\delta M/M \ll \delta V/V$, $\psi \approx \bar{\phi}$. Since the phase shift may be positive or negative, we get a total phase smearing of approximately $4\bar{\phi}$. It follows that

$$\frac{M}{\delta M} = \frac{\pi \Sigma n_i}{4\psi}. \tag{26.22}$$

Approximating $\Sigma\, \Delta V_{j-1}$ by $j\, \Delta V_m$ show that the resolving power is

$$R = \frac{M}{\delta M} \approx \frac{\pi \Sigma n_i}{4 \operatorname{arc\,cos}\{1 - (s_j/x_j)[1 + (V_B/j\, \Delta V_m)]\}}. \tag{26.23}$$

The construction and performance of such a mass spectrometer with a resolution $R \approx 200$ is described in detail in the literature.

26.3. Exercises

26.3.1.

Is it possible for electrostatic time-of-flight spectrometers to get ion transmission factors comparable to electromagnetic mass spectrometers?

26.3.2.

Try to design an electromagnetic instrument for similar mass range and resolution. Estimate the weight differences.

27. MASS SPECTROMETER FOR REACTOR FUEL RESEARCH*

27.1. Problem

Design a mass spectrometer capable of determining the isotopic content of uranium, thorium, and plutonium in irradiated fuel samples.

27.2. Solution

27.2.1. Introduction

Reactor fuel is subject to chemical reprocessing because an appreciable amount of fissionable material remains unused in present day reactor systems. Burnup of the fissile material originally introduced into a fresh reactor will in many cases be less than 50%. Even if conversion or breeding takes place from fertile into fissile materials, there will always remain some fissionable material left over in spent fuel. For reactor performance and behaviour considerations, as well as for economic reasons (namely to judge the economic feasibility of the reprocessing procedure), it is necessary to know at any instant the isotopic content of uranium and plutonium in spent fuel as precisely as possible. If a chemical solution of the dissolved fuel is available it is possible to get the necessary information with a mass spectrometer especially adapted to handle radioactive substances. To investigate solid fuel samples without chemical solubility, special ion sources have to be developed. An important consideration is the high α, β, and γ activity of spent fuel. An amount of 1 g of U^{235} fissioned results in an activity of about 3×10^{14} disintegrations/sec if the sample is prepared right after the fuel is taken out of the operating reactor. Special attention in mixed-type fuel must be given to the toxic effects of plutonium. Examine the health and safety literature for evidence of the danger of certain forms of plutonium if incorporated by human beings even in small amounts. Thus, two main aspects of the mass spectrometer to be dealt with are (1) proper sample preparations combined with the application of special ion sources and (2) glove-box-type operation.

* Problem 27 is by M. J. Higatsberger.

27.2.2. Experimental Methods

The mass spectroscopic investigation may be carried out with any glove-box-type spectrometer or with an instrument[1] capable of operation by remote control. For the analysis of liquid samples a triple-filament surface ionization source may be used. Describe the principle of operation and typical operating characteristics of such a source.[2] Qualitative determination of uranium, plutonium, and thorium in fuel samples can best be carried out by applying the isotope dilution method.[3] The precision of such measurements is essentially determined by the accuracy of the tracer concentration, by the mixing ratio, and by the mass spectrometrically determined ratio of the isotopes.

If x atoms of an element with n isotopes having abundances a_i where

$$\sum_{i=1}^{n} a_i = 1 \tag{27.1}$$

are mixed with y atoms of the same element with different isotope abundances, b_i, where

$$\sum_{i=1}^{n} b_i = 1, \tag{27.2}$$

show that the ratio c_i/c_k of the isotope i and to the isotope k in the mixture is given by

$$\frac{c_i}{c_k} = \frac{xa_i + yb_i}{xa_k + yb_k}. \tag{27.3}$$

If y is known, show that x is determined by

$$x = y \frac{(b_i/b_k) - (c_i/c_k)}{(c_i/c_k) - (a_i/a_k)} \frac{b_k}{a_k}. \tag{27.4}$$

27.2.3. Sample Preparation

Coated fuel particles are becoming more and more important in nuclear fuel technology. Normally they have a kernel with uranium or thorium carbide or oxide of about 500 μ in diameter and are coated in a fluidized

[1] M. J. Higatsberger, and F. P. Viehbock, "Electromagnetic Separation of Radioactive Isotopes." Springer, Vienna, 1961.

[2] W. L. Mead, ed., "Advances in Mass Spectrometry," Vol. 3. The Institute of Petroleum, London, 1966.

[3] M. L. Smith, ed., "Electromagnetically Enriched Isotopes and Mass Spectrometry." Butterworth, London, 1965.

bed by carbon through a cracking process at very high temperatures. The coatings are usually about 100 μ thick and ensure that the fission products, even under the most unfavourable conditions of chemical attack and high temperatures, do not penetrate the coating barrier.

Assuming that the fuel sample is available in the form of uranium–thorium carbide small particles, coated with pyrolitic carbon, mixed with graphite powder, and pressed to cylindrical compacts, the preparation technique is quite difficult because the known solvents do not attack the compact nor the fuel particles under normal conditions. The pyrolitic coating may be burned off by heating the sample in an oxygen stream for two hours at a temperature of 850°C. The resulting particles, containing only UC_2 and carbon coating in a 100 mg sample, may be dissolved in HNO_3, and thus the liquid sample can be applied to a mass spectrometer ion source in the usual way. Particles, containing (U, Zr)C and carbon coating, are heat treated at 850°C in an oxygen stream and rinsed with HNO_3, then H_2SO_4 is added and evaporated to dryness. Dissolving the particles in aqua regia then results in a usable sample. Finally, particles of $(U, Th)C_2$ with coatings of layers of C and SiC must be mechanically milled before fuming with H_2SO_4 and HNO_3 can take place. One finds even then that SiC or SiO_2 is undissolved.

If only the uranium isotope ratio is required a chlorination method avoiding any wet chemistry may be applied. In a silicate tube furnace the coated particles are burnt in an oxygen stream to remove the coating. Then they are mixed with graphite and chlorinated in a Cl_2 stream at 850°C. Instead of this two-step operation, a single-step technique can be used if the oxidation of the outer carbon layer is performed with CO_2 at temperatures near 850°C. A mixture of the gases of CO_2 and Cl_2 allows the removal of the coating and a volatilization of the uranium in the kernel at the same time.

27.2.4. Ion Source for Solid Sample Material

In order to restrict the handling and transportation of the highly radioactive material it is advantageous to construct an ion source specifically to analyze the particles. Use can be made of the fact that at very high temperatures (about 2500°C) the carbon coating cracks and then chlorination may be applied to dissolve the uranium in the kernel. Details of such a source are shown in Fig. 27.1. The source unit should be machined from one piece of graphite because of high temperature conditions. If several samples are to be analyzed, it is advisable to use a new source unit each time to avoid any possible memory effects.

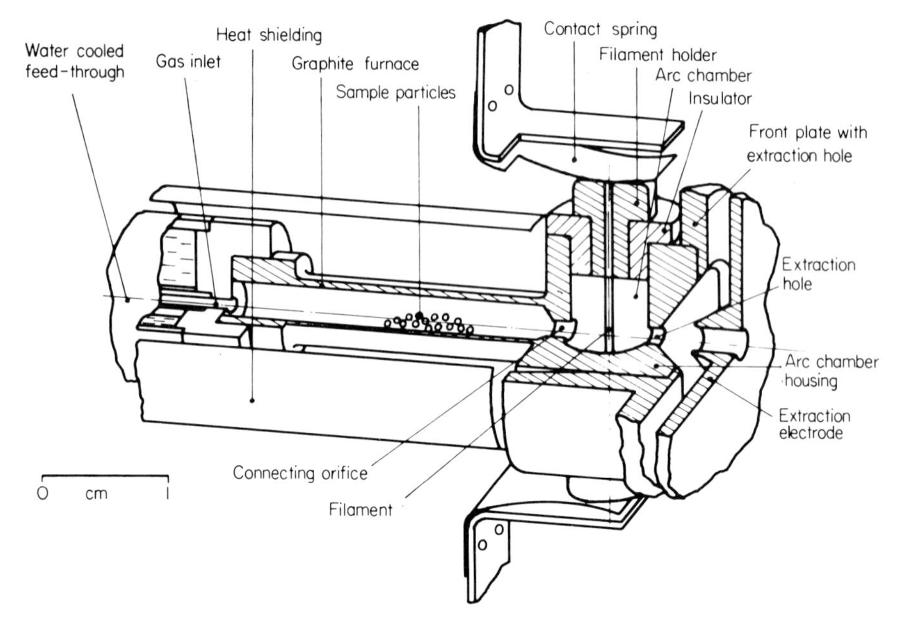

FIG. 27.1 Ion source insert for coated particles.

27.3. Exercises

27.3.1.

Calculate depletion and burnup of U^{235} from measuring the ratio U^{236}/U^{235} before and after irradiation. Discuss the accuracy of the result.

27.3.2.

Outline a procedure for measuring, by means of mass spectroscopy, the most important fission products for yield determination.

28. ISOTOPE SEPARATOR*

28.1. Problem

Design a multipurpose β-γ-radioactive isotope separator with an output of at least 10 mg/h for the most abundant separated isotope. What provisions must be made to collect 10 mg of radioactive Xe^{131}?

28.2. Solution

28.2.1. Introduction

The most intensive β-γ-radioactive isotopes are found in fission products. As an average mass value an atomic weight of $A = 100$ may be assumed. If 10 mg/h of a single isotope of mass 100 is to be collected, show that a minimum collector current of roughly 3 mA is needed. (Assume each ion collected is singly charged.) For Xe^{131} separation it is important to consider equilibrium conditions; Xe^{131}, with a half-life of 12 days, is produced by β decay of I^{131}, which has a half-life of 8.08 days. Xenon isotopes ranging from mass 129 up to 143 are found in spent nuclear fuel. Normally, it is only necessary to regard the isotopes up to mass 136 since the heavier isotopes are very shortlived. A possible means to get the required 10 mg of pure Xe^{131} would be by a chemical oxidation process. Iodine can be separated out of fission products in liquid solution. From the iodine isotopes the numerous xenon isotopes are produced with intensities related to the intensity and decay constant of the parent isotopes. If a machine is constructed capable of delivering up to 10 mA collector current, we need only collect 1 mA of Xe^{131} current in a liquid–air cooled collector pocket for approximately 2 h. Means must be provided for recovering the gaseous Xe^{131} from the collector pocket by a mild heating process, and pumping it into an evacuated vessel. When separating radioactive isotopes, it is very desirable to have a large enrichment factor as well as a high luminosity of the machine. Because of the greater mass dispersion and the two-directional focusing property an inhomogeneous magnetic field is used

* Problem 28 is by M. J. Higatsberger.

for a mass separation field.[1] When working with radioactive isotopes, it is advantageous for handling purposes to have the ion source and the collector outside the magnetic field. Therefore, a magnetic sector field is required. The ion optical properties of inhomogeneous magnetic sector fields, including the trajectories out of the medium plane, have been calculated by Tasman *et al.*[2,3]

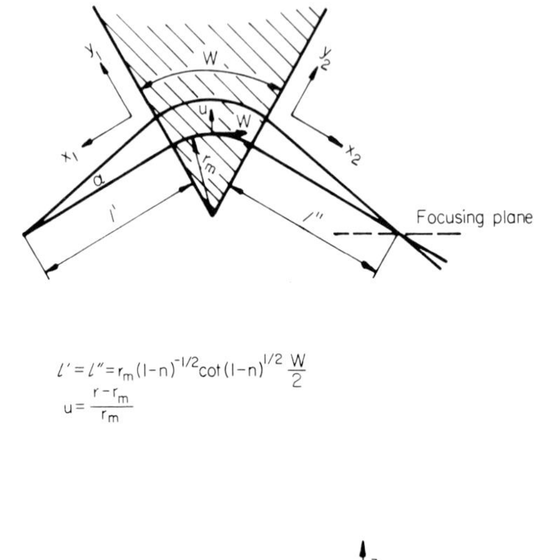

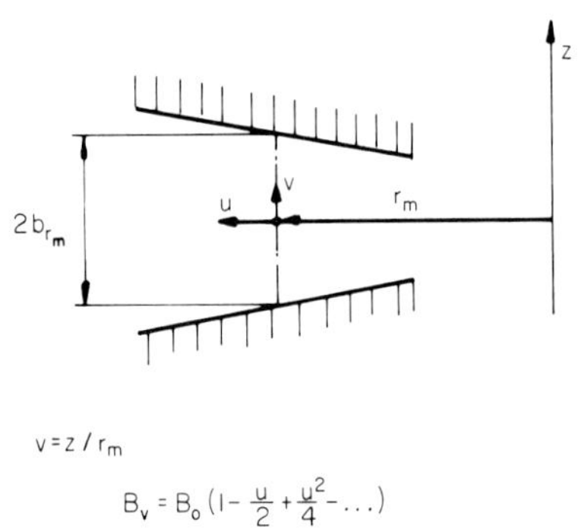

FIG. 28.1. Coordinate system of the magnetic field.

[1] N. Svartholm, and K. Siegbahn, *Rev. Sci. Instr.* **19,** 594 (1948).
[2] H. A. Tasman, and A. J. H. Boerboom, Z. Naturforsch. **14a,** 121 (1959). H. Wachsmuth, A. J. H. Boerboom, and H. A. Tasman, Z. Naturforsche. **14a,** 818 (1959).
[3] M. J. Higatsberger, and F. P. Viehbock, "Electromagnetic Sep. of Radioact. Isotopes." Springer, Wien, 1961.

28.2.2. The Magnetic Field

The mass dispersion coefficient of an inhomogeneous field is given by

$$D = \frac{r_m}{(1 - n)} \frac{\Delta m}{m} \tag{28.1}$$

where

$$n = \frac{\delta B_z}{\delta r} \frac{r}{B} \tag{28.2}$$

is the logarithmic field gradient and B_z the z component of the magnetic strength. In Eq. (28.1) r_m is the radius of the central trajectory. The coordinate system is shown in Fig. 28.1. If we choose the central radius $r_m = 1000$ mm and the inhomogeneity $n = \frac{1}{2}$, what is the spatial separation due to dispersion for an atomic mass 240 and for $\Delta m = 1$? Show that the object distance, which in this case is equal to the image distance, is given by

$$\frac{l'}{r_m} = \frac{l''}{r_m} = (1 - n)^{-1/2} \cot (1 - n)^{1/2} \frac{W}{2}. \tag{28.3}$$

If we want to have focusing in the z direction and second order focusing in the radial direction too, show that the sector angle of the magnetic field must be $W = 169°42'$. The magnetic field can be realized by conical pole faces. Make a graph of the axial component of the magnetic field strength (B_r/B_{r_m}) in the central plane. What are the object and image distances? Because of the fringing fields at the boundaries of the magnetic field, additional deflections of the ion beam will occur at the entrance and at the exit of the field. These deflections will be about 5° on each side. Thus, the total deflection amounts to about 180°. In Fig. 28.2 the ion trajectories are drawn for a radial opening angle of $a_r = \pm 5°$ and a mass difference of $\Delta m = \pm 5\%$. In the z direction the height of the object slit is 50 mm and the axial opening angle $a_z = 1\frac{1}{2}°$. Under these conditions, show that the maximum width of the ion trajectories in the central plane will occur at a sector angle of $w = 113°$. What is this width? What is the width of the ion beam at the entrance and at the exit of the magnetic field? Show that the beam has its maximum height of 112 mm at a sector angle of $w = 78.5°$. What is the beam height at the entrance and at the exit of the magnetic field? For a momentum spread of $\beta = 10^{-4}$ and under the conditions mentioned above, evaluate the central-mass-image broadenings due to the following first and second order aberrations:

(a) first order chromatic aberration, Δy_β;

(b) second order radial angular aberration, $\Delta y_{a_r 2}$;
(c) mixed aberration, $\Delta y_{a_r \beta}$;
(d) second order chromatic aberration, Δy_{β^2};
(e) second order axial angular aberration, $\Delta y_{a_z 2}$;
(*f*) agammatism, $\Delta y_{a_z \delta}$;
(g) image curvature, Δy_{δ^2}.

Show that the radius of curvature inside the magnet is $R_{im} = -650$ mm. The negative sign of R_{im} means that the center of curvature has a negative

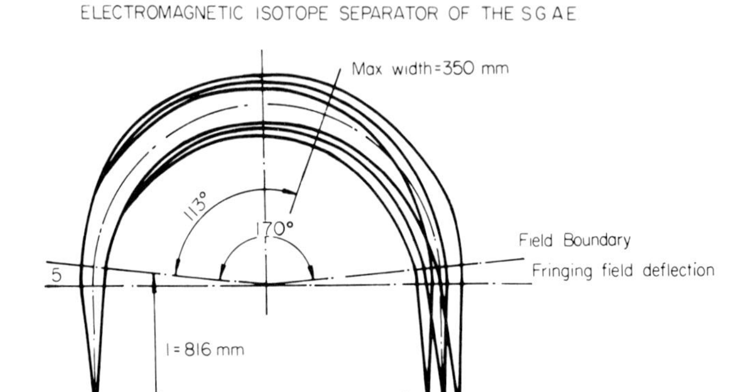

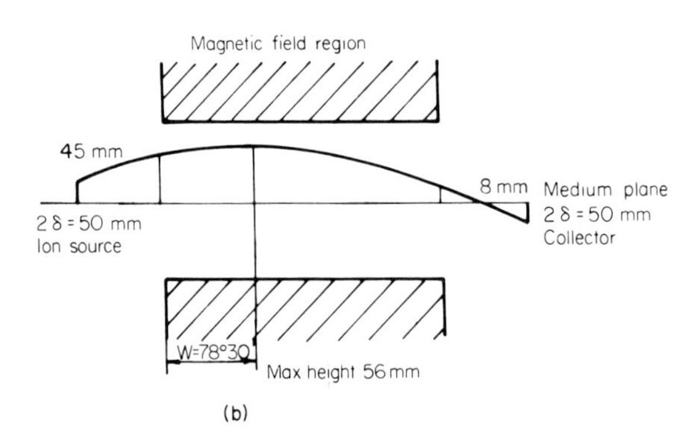

FIG. 28.2. Schematic ion trajectories of the electromagnetic isotope separator of the S.G.A.E.: (a) in the medium plane, where $r_m = 1000$, $\alpha_m = \pm 5$, $\Delta_m = \pm 5\%$; (b) in the z direction.

y coordinate. Due to the fringing fields, show that there is an additional image curvature with the radius $R_{fr} = -500$ mm. What is the total image radius of curvature?

Preferably, the magnet should have the form of a "C"; the yoke should be made of soft cast iron and the pole pieces of wrought iron. If the maximum field strength at medium plane is 5000 Oe, how many ampere-turns must the coil have? If there are 2500 turns, what is the maximum current? What is the maximum power consumption of the coil? Is cooling required? Estimate the total weight of the magnet including the coil.

28.2.3. The Ion Source

Details of the design of a source can be seen in Fig. 28.3. The ion source we shall use is a conventional arc-discharge source with an electron

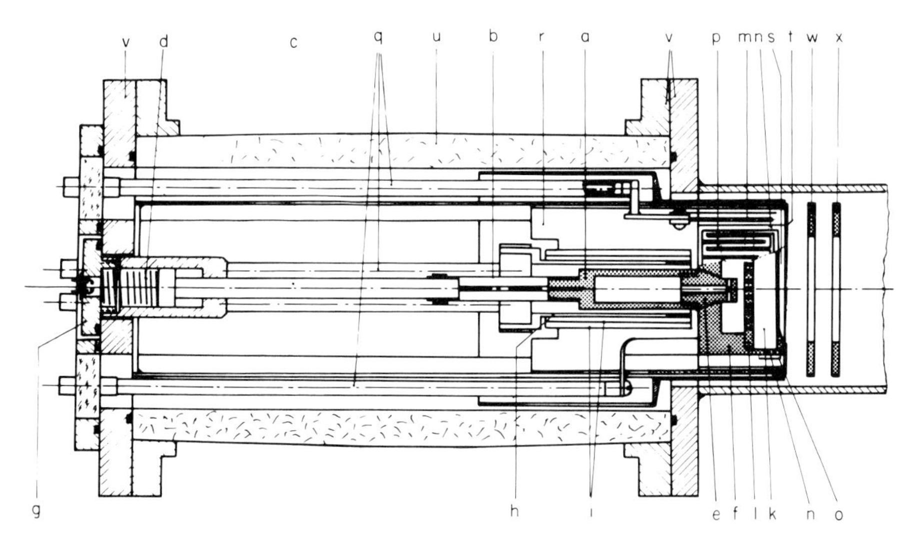

FIG. 28.3. Ion source.

collimating magnetic field perpendicular to the ion beam. The charge material will be loaded in the graphite furnace (a). The furnace is made from pure reactor graphite and if desired, the chemically prepared charge material can be irradiated in the reactor together with the furnace. By a thin molybdenum tube (b), the furnace (a) is connected with a mobile stainless steel tube (c). By means of a spring (d) the conical front piece (e) of the vapor furnace (a) is pressed against the conical hole in the arc-chamber box (f) which is also made of graphite. The vapor furnace can be loaded and unloaded with the aid of four screw bolts on the flange (g). The vapor furnace (*a*) is heated up to temperatures above 1000°C by means of a cylindrical graphite heater (*h*); this heater consists of a cylindrical graphite tube with a wall thickness of 2 mm which has six longitudinal slits along its circumference. The power consumption of the furnace will be 2 kW. Two cylindrical heat radiation shields (*i*) of molybdenum are arranged around the heater (*h*) to suppress the dissipation of heat. The vapor atoms entering the arc chamber (*k*) through a hole in the conical front piece (*e*) of the furnace are distributed over the entire chamber by means of a graphite distributor plate (*l*). The electrons emitted from a tungsten filament (*m*) and collimated by a magnetic field enter the arc chamber (*k*) through the hole (*n*). A repeller electrode (*o*) is provided to keep the electrons oscillating in the arc chamber. Additional heating of the arc-chamber box (*f*) is possible by means of the two graphite heater plates (*p*). The conducting tubes (*q*) of the filaments as well as of the furnace and of the chamber heaters are water cooled. The arc chamber and vapor furnace region of the ion source are surrounded by stainless steel shieldings (*r*). In front of the ion extraction slits (*s*) there is a molybdenum heat shield (*t*) in the form of a Pierce lens. The ion source unit is at a potential of + 50 kV above ground and is insulated by a porcelain tube (*u*). The flanges (*v*) both at high and at ground potentials are water cooled. The ion extraction electrode (*w*) is at a potential up to − 30 kV and is situated at a distance of 15 mm in front of the source slit. This negative electrode is water cooled and can be manipulated from the outside of the vacuum chamber in order to facilitate the calibration of the ion beam. The distance between the negative electrode (*w*) and the grounded electrode (*x*) is 10 mm. The distance between the ion source shielding (*r*) and the inner wall of the vacuum chamber is quite small. This ion source is primarily designed to be used for solid charge material, but a gas inlet system can be provided.

28.3. Exercises

28.3.1.

What are the fundamental differences between isotope separators for stable isotopes, for separation of α activities, and for separation of β–γ activities?

28.3.2.

List some of the advantages of a high intensity isotope separator attached to an accelerator. What problems must be dealt with if the ion source and the collector are to be designed to deal with currents in the range of amperes?

29. HALF-LIFE OF N^{13}*

29.1. Problem

Measure the half-life of N^{13} with an accuracy of one part in a thousand.

29.2. Solution

The accuracy of half-life measurements is limited by the following factors: (1) counting statistics, (2) purely instrumental errors, and (3) contaminations in the source.

In order to reach an error of 10^{-3} it is necessary to count at least 10^6 events. This very simple statistical consideration is sufficient for the specific-activity method (cf. Chapter 2.6.1.4. of Vol. 5B, this treatise). When following the decay of a given source directly (cf. Chapter 2.6.1.3, this treatise), one may expect, however, to need to count at least the same order of magnitude of events, as can be seen from a decay curve measurement with only two points, at an interval of one half-life. The half-life of N^{13} is about 10 min. This rough value can be taken from any compilation. In order to count 10^6 events within a half-life, an activity dN/dt of at least $dN/dt = 10^6/10$ min. $\approx 1.7 \times 10^3$/sec ≈ 0.5 mC is needed. On the other hand, it is easy to produce activities exceeding 100 mC, and to measure them with a suitable apparatus. This means that counting statistics need not limit the accuracy of the half-life measurement. Indeed, in practice, equipment with a detection probability much smaller than unity can be used, and more care can be taken to reduce other sources of error.

As N^{13} has a short half-life in comparison with, say, the mean time a physicist is working in one laboratory, it is convenient to follow the decay directly. The other method which may also be used, namely, the specific-activity method, is not as suitable. It does not offer any advantage: if it is *once* possible to measure the counting rate accurately enough, it will be possible to do it *again* with the same apparatus. In this manner the decay curve can be obtained. On the other hand, it is impossible to measure the activity with an accuracy of one part in a thousand. Note that those effects which reduce the precision of an activity determination, such as self-

* Problem 29 is by H. Daniel and W. Gentner.

absorption, or wall effects in a Geiger–Müller or a proportional counter, cancel when the decay curve is taken. Even more serious difficulties are to be expected in the determination of the number of radioactive nuclei per unit weight.

After having decided to follow the decay directly the choice of a detector has to be made. The half-life is inconveniently short for an ionization chamber or a calorimeter. Geiger counters can not provide the high counting rates needed. A scintillation counter or a proportional counter is suitable. For brevity, we will consider only a scintillation counter in what follows. The discussion for a proportional counter will be essentially the same. Solid state counters have not yet been examined extensively enough to decide their suitability.

All the phototubes used in scintillation counters show a gain which varies with the counting rate. This variation depends strongly on the tube type. Since N^{13} has a continuous β^+ spectrum which extends from the maximum energy (1.19 MeV) down to zero, the counting rate will depend on the discrimination level, if the positrons are counted directly. Of course, a discrimination level of zero can not be used because of the thermal noise in the phototube and the associated circuitry. The pulse-height spectrum of the annihilation radiation, as given by a NaI scintillation counter, also shows a continuous distribution. Therefore, the same discrimination difficulties will arise. Because of the lower background provided by the thin anthracene or plastic scintillation counters which are used for the detection of particles, it is more advantageous to count the positrons directly rather than their annihilation quanta.

With magnetic energy selection of the positrons, one gets a *line* spectrum when counting the positrons directly, provided that the scintillation crystal is thick enough to stop the positrons. If it is thinner, one gets instead a more or less continuous pulse-height distribution. In either case, however, it is possible to have a pulse-height spectrum which shows a low intensity between the peak and the noise region. When discriminating at this intensity minimum, even larger gain variations, which may arise from counting rate variations, do not substantially influence the detection probability.

Almost any type of β-ray spectrometer is suitable as an energy selection instrument or monochromator. The reader is referred to Chapter 2.2.1.1 of Volume 5, Part A, of this treatise. In addition it is not necessary to use a thin source, the source thickness must only be kept constant during a run. If there is no need to discriminate against a contamination within the source, it will be advantageous to set the spectrometer on the maximum of the β-ray spectrum in order to minimize the effect of small drifts in the momentum setting.

The detection probability of a magnetic spectrometer set at the maximum of a not too soft continuous β spectrum is roughly the product of transmission T and resolution η (η = ratio of full width at half-maximum to the center momentum of a line).

As the half-life is to be measured within one part in a thousand it is necessary to have a clock with at least this precision. Not every stop watch is accurate enough. An error of one part in a thousand means an error of 1.5 min/day. The power line frequency is usually not constant enough to serve as a time standard.

N^{13} can be produced by the deutron bombardment of carbon in a cyclotron according to the reaction $C^{12}(d, n)N^{13}$. It is easy to obtain specific activities of the order of 1 Ci per gram carbon. No disturbing background activities arise from the deuteron bombardment of pure carbon; therefore no chemical separation is needed. It may, however, happen that radioactive nitrogen escapes from the source. This possibility should be checked.

It is advantageous to start with a fairly high counting rate and to follow the decay of the source as long as possible. The first points serve as a check of the dead time correction, while the last points prove the purity of the source. The background (without source) should be measured immediately before and after taking the decay curve.

Before starting the analysis of the decay curve, the dead time correction must be applied to the counting rate whenever this correction is not negligible. The counting rate in the scintillation counter consists of two parts: the part due to the radioactivity in the source, and the part due to the background which comes from thermal noise, cosmic ray events, etc. As the background is kept small in our experiment because of the high discrimination level in the scintillation counter, it will be sufficient to assume a constant background value. This backgound is subtracted from the counting rate corrected for dead time losses. As the third step, the counting rate due to contaminations (if any) in the source is subtracted by extrapolating back the final part of the decay curve. A more elaborate procedure is not necessary in our case, because of the purity of the source. The remaining counting rate should be due only to N^{13} and should show a purely exponential decay. This can be roughly checked by plotting on semilogarithmic paper. For a detailed evaluation, however, this procedure is too crude. The accuracy of the graphpaper grid is not sufficient. Instead, the value taken from the semilogarithmic plot serves as an approximate value. By plotting the difference between the measured counting rate corrected for dead time losses and background and the rate calculated with the approximate half-life the scale is considerably enlarged. One may also perform a least-squares fit and compute the standard deviation. In cases such as the present where it is easy to obtain good statistics, this standard deviation does not tell too much about the overall precision of the measurement. A better way to determine the experimental precision is to perform several (not too few) independent runs, and to compare the variations in the results. This, however, does not eliminate certain systematic errors. The large number of cases where different authors have reported remarkably different half-lifes but each with a very small quoted error for the same decay, shows us how easy it is to underestimate systematic errors.

30. NUCLEAR COUNTING EXPERIMENTS*

30.1. Problem

What can you state about intensity, background, and resolution of nuclear counting experiments?

30.2. Solution

30.2.1. Introduction

Attention must be given to the following considerations when counting nuclear particles or quanta: Is, in a given time interval, a certain type of measurement feasible at all, and if some degree of freedom is still open in a typical arrangement, what free parameters are most suited for a definite kind of measurement? For a counting experiment, the law of statistical counting is:

$$\Delta N = \sqrt{N} \tag{30.1}$$

where ΔN is the standard error of the counting measurement related to the total number of N counts.[1] What is the standard error of the sum of several independent counts $(N_1 + N_2 + \ldots)$?

30.2.2. Counting Rate Measurements in the Presence of Backgrounds

The "quantity of information" J_S for a quantity S is defined as

$$J_S = \left(\frac{S}{\Delta(S + B)}\right)^2. \tag{30.2}$$

S is the number of true counts and $\Delta(S + B)$ the total standard error, including the effect of background. This quantity has certain additive properties, e.g., proportionality to the counting time and to the counting rate. These proportionalities can be better seen when J_S is written in a form valid for the cases to be treated below:

$$J_S = SW, \qquad 0 \leqslant W \leqslant 1. \tag{30.3}$$

[1] G. Quittner, Intensity, background and resolution width in nuclear counting experiments, *Nucl. Instr. and Methods* **31,** 61–67 (1964).

* Problem 30 is by M. J. Higatsberger.

When comparing particular *measurements*, S is the total number of true counts and J_S is the total amount of information about the true counting rate. When comparing *instruments*, S represents the number of true counts per time unit, i.e., the true counting *rate* and J_S is then accordingly the *rate* of accumulation of information about the true counting rate; W, being independent of counting time, is the same in both cases.

W can be expressed as the ratio

$$W = \frac{\Delta^2 S}{\Delta^2(S + B)} \tag{30.4}$$

of the variances of an associated "ideal" measurement (with the same number of true counts but without background) and of the actual measurement. In the limiting case of the "ideal" measurement, $W = 1$, and according to Eq. (30.3) the amount of information is numerically equal to the number of true counts, whereas in the nonideal measurements J_S is less than unity by a factor W. The factor W is therefore a measure of the efficiency of conversion of (true) counts into information. Equivalently, we can say that a particular measurement (with background) has delivered J_S "effective counts," because it has given the same information as an ideal measurement with J_S counts. If the mean value of the background is known in advance, show that W can be expressed in terms of $\beta = S/B$, the ratio of the true to background counts according to

$$W = W_1(\beta) = \frac{1}{1 + (1/\beta)}. \tag{30.5}$$

If the background is not known, a certain percentage, $1 - \alpha$, of the whole time must be used for a "sample-out" measurement to determine the background and only a fraction α remains for the "sample-in" measurement Show that

$$W = W(\alpha, \beta) = \frac{\alpha}{1 + 1/[\beta(1 - \alpha)]}. \tag{30.6}$$

The arbitrary value α may be chosen in a way to render $W(\alpha, \beta)$ maximum, then

$$W = W_{\max}(\beta) = W(\alpha_{\text{opt}}(\beta), \beta). \tag{30.7}$$

An important special case of Eq. (30.6) is

$$W(\tfrac{1}{2}, \beta) = \frac{1}{2(1 + 2/\beta)}. \tag{30.8}$$

Make a graph of the functional forms of $W_1(\beta)$, $W(\frac{1}{2}, \beta)$, and $W_{\max}(\beta)$

for $0 \leqslant \beta \leqslant 10$. Show that for small β:

$$W_1(\beta) \approx \beta \tag{30.9}$$

and

$$W_{max}(\beta) \approx W(\tfrac{1}{2}, \beta) \approx \tfrac{1}{4}\beta \tag{30.10}$$

which are useful for considerations of the limiting case of high background. Show that $W(\frac{1}{2}, \beta)$ is a good approximation to $W_{max}(\beta)$ up to $\beta \approx 5$. Therefore $W(\frac{1}{2}, \beta)$ can be used instead of W_{max} in this range, with the advantage of greater simplicity.

30.2.3. Examples

30.2.3.1. Evaluation of Counting Situations. In Fig. 30.1 two peaks are shown, No. I with 100 counts and No. II with 5 counts. The S/B ratios

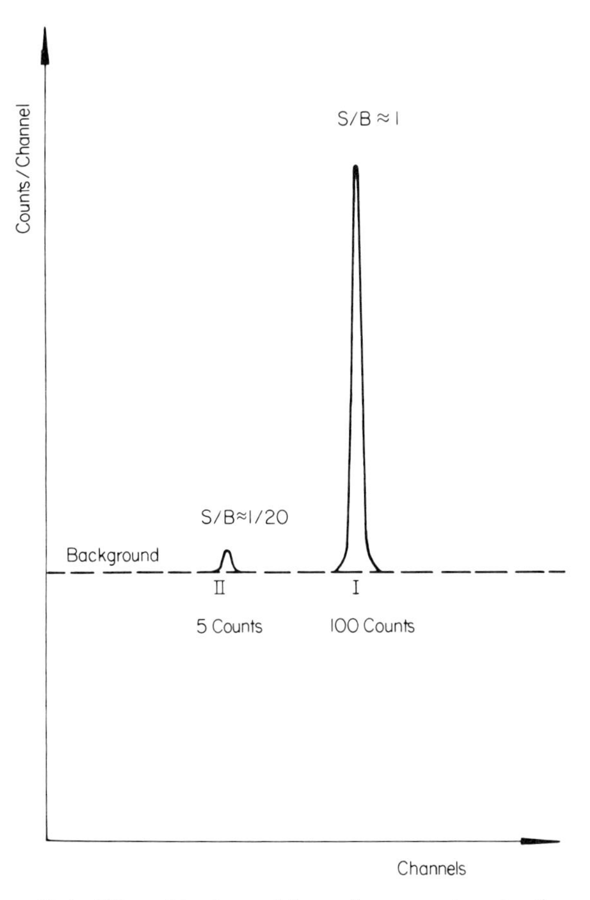

Fig. 30.1. Effect of backgound for a given counting situation.

are 1 and 0.05, respectively. What are the effective counts for W_1, and for $W(\frac{1}{2}, \beta)$?

30.2.3.2. Thickness of Crystal Neutron Filters.[2] It may be advantageous to sacrifice useful particles if a sufficiently large decrease of background is achieved. The limit of useful increase of thickness of such a filter is given, when sources of background become dominant which are not affected by

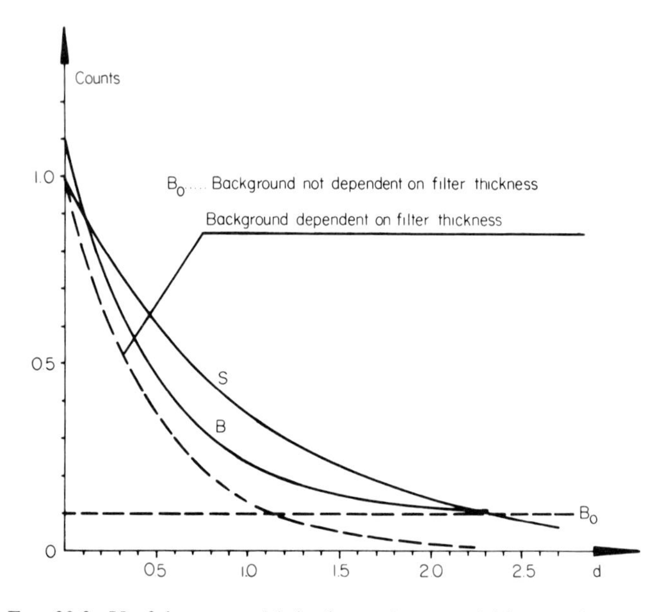

FIG. 30.2. Useful counts with backgound versus thickness of neutron filters.

the filter (B_0 in Fig. 30.2). The optimum thickness, however, may be found by evaluation of Eq. (30.3), if S and B are known as functions of the thickness of the filter. What is the optimum thickness for the data given in Fig. 30.2.

30.2.3.3. Filling Pressure of BF_3 Counters. The useful counting rate and the background counting rate of BF_3 counters depend, under certain conditions, on the filling pressure p in a manner shown in Fig. 30.3(a): these conditions are fulfilled rather well in BF_3 tubes placed perpendicular to a neutron beam and, for the "parallel" arrangement, for tubes without a "dead volume." Fig. 30.3(b) shows the optimal counting efficiency y as a function of A, the S/B ratio at small pressures. Under what condition is

[2] B. N. Brockhouse, Methods for neutron spectrometry, Proc. Symp. Inelastic Scattering Solids and Liquids, Vienna, 11–14 October, 1960, IAEA, (1961).

a large filling pressure and high counting efficiency not advantageous? How is the optimum counting efficiency related to the relative background [B in Fig. 30.3(a)]? With BF_3 tubes having a considerable dead volume, used in the parallel arrangement, the dependence of S on P is of the sort indicated by the dashed line of Fig. 30.3(a). What filling pressures are favoured under these conditions?

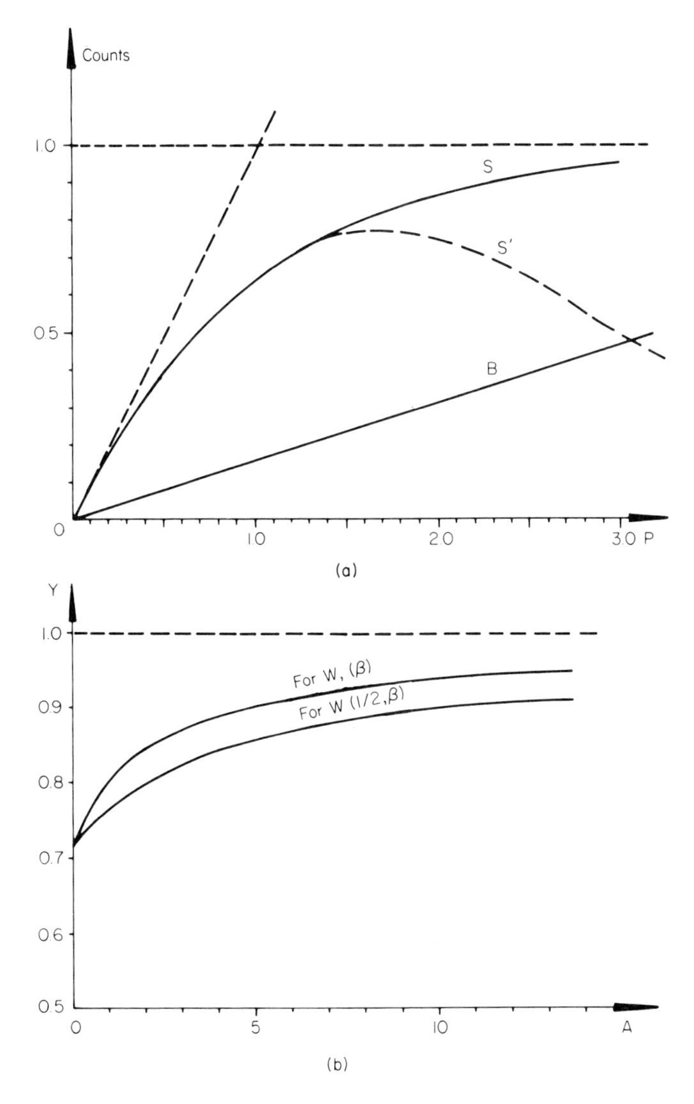

FIG. 30.3. Useful counts for BF_3 counters.

30.2.4. Location Measurements

In this class of measurements, a distribution of counting rates Y versus some physical variable X is evaluated to find some definite value X_0 of the physical variable. If the distribution consists of a relatively large number of counts near X_0, and a relatively small number at other values of X, one speaks of a "peak." Examples are energy determination of γ lines or dispersion relation measurements by inelastic scattering of neutrons.

As an example, the evaluation of the gain stability of a γ-scintillation spectrometer will be considered. We define the "location uncertainty" δx to be of such a magnitude that the change of the "true value" of Y due to the shift δx is equal to the statistical uncertainty δY.

$$\delta x = \delta Y \frac{dx}{dY}, \qquad Y = yD, \qquad \delta^2 Y = \frac{Y}{W(x)}$$

where y is the x density of counting rate and D is the channel width. Then we define

$$\Delta J_x = \frac{1}{\delta^2 x} = \frac{D(dy/dx)^2 W(x)}{y} \tag{30.11}$$

as the "amount of location information given by a channel of width D." Correspondingly,

$$i_x(x) = \frac{(dy/dx)^2 W(x)}{y} \tag{30.12}$$

is the x density of location information, and

$$J_x = \int_{\text{peak}} W(x) \frac{(dy/dx)^2}{y}\, dx \tag{30.13}$$

is the total amount of location information of the peak. For quick estimates of J_x for peaks of the same shape but different "heights" and widths, show that Eq. (30.13) takes the following form:

$$J_x = C_1 \bar{W} \frac{y_0}{H} = \frac{C_2}{H^2} \bar{W} \int y(x)\, dx \tag{30.14}$$

where C_1 and C_2 are shape factors, y_0 is the counting rate density at the top of the peak, and H the full width at half-maximum. Show that $C_1 = 5.88$ and $C_2 = 5.55$ for Gaussian peaks. At low background ($W \approx 1$), under what conditions are small and narrow resolution peaks equivalent to larger but broader peaks? Narrower peaks have obvious advantages, e.g., better resolution and smaller systematic uncertainty. Under practical conditions, however, these advantages are limited by background because if narrow resolution peaks are too small $W \approx 1$ no longer holds.

30.2.4.1. Examples

1. A formula used by Brockhouse[3] in the evaluation of statistical accuracy of peak location in his treatment on the measurement of the Kohn effect by neutron spectrometry is

$$\delta = \frac{\sqrt{N}}{\sqrt{n}} \frac{dv}{dN} \tag{30.15}$$

where v is the frequency of a normal mode of vibration, N is the mean number of counts per measuring point, and n is the number of measuring points. Starting with Eq. (30.11), taking $W \approx 1$, using $N = Dy$, and taking dN/dx constant, develop Brockhouse's formula. (The formula is evaluated over the steepest portion of the resolution peak and this portion can be approximated by a straight line.) It is interesting to note that in this experiment the final statistical uncertainty of peak location was about $\frac{1}{30}$ of the full width at half-maximum of the resolution peak.

2. Choice of flight-path (L) in time-of-flight neutron spectrometers. Two cases are considered:

(a) The detector area is constant. Then the integral intensity Cy_0H is proportional to $1/L^2$.

(b) The detector angle is constant: $Cy_0H = \text{const}$.

With regard to the background conditions only the extreme cases will be treated: low background ($\bar{W} \approx$) and high background ($\bar{W} \approx y_0/B$; see Eqs. (30.9) and (30.10). B is the background level in the detector [not to be confused with the spatial background density $b(L)$; in fact, B is proportional to $b(L)$ times the detector area]; $H(L)$ is the dependence of the spread H of the time-of-flight, s, on the distance L between source and detector; $b(L)$ is the spatial background density. We first evaluate Eq. (30.14) for $x = s$ (the time of flight) and then convert to J_v, the velocity location information. Show that $J_v = L^2 J_s$. (We assume therefore that we are interested in the location of a velocity.) Evaluate J_s and J_v in terms of L, $H(L)$, $b(L)$, and a proportionality constant C for the two cases indicated above for both low and high backgrounds.

If it is assumed that the spatial background level $b(L)$ is constant in space, and that $H(L)$ is a monotonically increasing function of L, beginning with a dependence $H(L) \approx \text{const}$ in the neighborhood of the source and ending with $H(L) \approx L$ at large distances, is there an advantage in taking

[3] B. N. Brockhouse *et al.*, *Phys. Rev.* **128**, 1099 (1962).

long flight paths? Under what conditions is the shortest possible flight path advantageous? Statistical accuracy is not the only factor determining the usefulness of a measurement. In the present case, questions of resolution, mean velocity of neutrons, repetition rate, size of sample and detector, etc., must be taken into account before a final conclusion on the choice of L is reached.

30.2.5. Intensity Distribution Measurements

The simplest example is the comparison of the intensities of two γ lines, as contrasted to the determination of their energies in the "location" type of measurements. The generalization of this comparison as, for example, in the measurements of cross sections as a function of energy, is more common. The measurement has to be extended over a certain range $x_1 \leqslant x \leqslant x_2$ of the variable x. The function $y = f(x)$ is, in practice, sampled in a finite number n of points which must be chosen by the experimenter. The range of the variable x and the number n determine the "sampling interval" Δx. The experimenter has to decide on the resolution width H of his instrument. A relationship for the apparatus at hand between the integral intensity of the resolution peak (corrected for background) and its half-width might take the form:

$$J_y = \int y(x)W(x)\,dx = \text{const } H^{\nu} \tag{30.16}$$

where ν may be between $1 \leqslant \nu \leqslant 2$ for good background conditions and $1 \leqslant \nu \leqslant 3$ for bad conditions. As long as $\nu > 1$, the measurement, over a certain interval Δx of the function $f(x)$ [or more precisely of the mean value of $f(x)$ over this interval] is better performed in a single measurement with a resolution width H, than by two successive measurements with a width $H/2$ and finally averaging the result. Why? But this means the resolution width should be as large as possible. The limit is given by two considerations.

1. One is in general interested in measuring as many "details" of the function $f(x)$ as possible. There is, however, a definite relationship between the sampling step Δx and the amount of detail of a measured function: this is an analog of the "sampling theorem" of communication theory[4]

$$\Delta x = \tfrac{1}{2} f_{\max} \tag{30.17}$$

where $f_{\max}$ is the frequency of the highest harmonic of the Fourier spectrum of $f(x)$. The cutoff frequency $f_{\max}$ is, in the case of a measurement, given

[4] J. R. Ragazzini, and G. F. Franklin, "Sampled Data Control System," McGraw–Hill, New York, 1958.

by the requirement that the amplitude of the highest harmonic to be taken into account is of the same order as the "noise level" determined by the final uncertainty Δy of a single measurement and the number n. It is here that the "frequency spectrum" (a relationship which is determined by the nature of the samples under investigation and not by measuring apparatus) enters.

Klepikov and Sokolov[5] have given a theory of these problems for the case of general statistical uncertainty δ (their use of orthogonal polynomials instead of Fourier harmonics does not alter the essential features of the problem). According to them, the "noise level" Σ is in the order of

$$\Sigma \approx \frac{1}{m} (\Delta y \text{ of a single measuring point}).$$

The number m should be such that

$$q \equiv n - m \approx 2 \text{ to } 3. \tag{30.18}$$

2. All these considerations are valid in the case where the measured values are independent of each other and this is certainly true if H is "sufficiently small" compared with Δx. The more general theory for the case of interdependent measuring points is apparently even more complicated, but one can avoid the complications by requiring that the magnitude of the resolution correction, (which, e.g., for a Gaussian resolution peak of width H neglecting higher derivatives of $f(x)$, is

$$d(x) = 0.09(d^2y/dx^2)H^2, \tag{30.19}$$

should be of the order of the width of the error strip of the result without resolution correction. The latter is, according to Klepikov and Sokolov, of the order $(m/n)^{1/2}\ \Delta y$, so that our requirement is

$$0.09(d^2y/dx^2)H^2 \approx \left(\frac{m}{n}\right)^{1/2} \Delta y. \tag{30.20}$$

Given the total measuring time T, the relationship (30.16) of the apparatus, and the frequency spectrum of the functions $f(x)$, one is, in principle, in a position to calculate the n and H of a rational experimental design. Figure 30.4 shows the pertinent logical relationships for the calculations of q of Eq. (30.18). Starting with trial values of n and H, one can see whether conditions (30.18) and (30.20) are satisfied. Needless to say, in practical cases the necessary information (especially on the harmonics

[5] N. P. Klepikov, and D. N. Sokolov, "Analysis and Planning of Experiments by the Method of Maximum Likelihood." Pergamon Press, Oxford Univ. Press, London and New York, 1962.

spectrum) is incomplete or only qualitatively known at the beginning, but may be accumulated during the course of measurements. It may then be worthwhile to reexamine the initial choices of n and H.

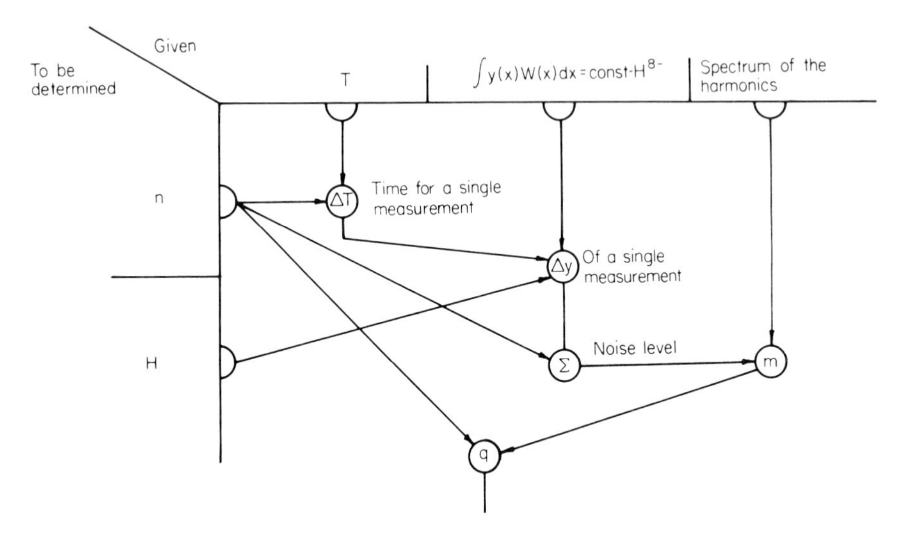

FIG. 30.4. Determination of q according to Eq. (3.18).

30.2.6. The Matching Problem

The rate of decrease of statistical uncertainty may be calculated by formulas such as (30.3) or (30.13). Among the sources of final uncertainty the statistical uncertainty has the unique feature that it is decreased by measuring for a longer period of time: in principle it can be made arbitrarily small. In practical measurements, the rate of decrease of statistical uncertainty must be considered in relation to the measuring time one is willing to spend for a particular measurement, and to other sources of uncertainty not depending on time, to get a reasonable balance between them. By sharpening or widening the resolution of the apparatus one can often alter the ratio between the statistical uncertainty and other error sources. In this respect there is a marked difference between relative measurements (i.e., measurement of all kinds of "shifts") and absolute measurements. An example of an essentially relative measurement is the measurement by Brockhouse cited above. Here we have seen that the statistical accuracy may be much smaller than the peak width. On the other hand if one had to measure a neutron velocity v absolutely, i.e., as a ratio s/L, then one would be forced to take a flight path long enough so that

$\Delta L/L$ and $\Delta s/s$ are compatible with the statistical accuracy attainable within a given measuring time T; the corresponding flight path length would then be different from that minimizing the statistical $\delta v/v$.

Generally, one has to consider all the theoretical and experimental information about such a problem and the apparatus, and match the parts of the apparatus to the whole apparatus and to the problem. A beautiful example of matching considerations for a neutron time-of-flight spectrometer is given in a paper by K. E. Larsson *et al.*[6]

30.3. Exercise

The foregoing considerations hold for simple locating measurements and intensity distribution measurements. Elaborate in more detail the relation of the statistical uncertainty to other possible sources of errors.

[6] K. E. Larsson *et al.*, *Arkiv Fysik* **16**/**19,** 199 (1959).

31. GAMMA SPECTROSCOPY*

31.1. Problem

Design a high resolution semiconductor Compton spectrometer for nondestructive burnup determination of nuclear fuel and outline the basic physics involved; list some of the most important γ-active fission products for use in burnup and burnup history determination and show the potential of such a device.

31.2. Solution

31.2.1. Introduction

The cumulative yield, characteristic γ-ray emission energies, decay schemes, and decay constants for most of the fission products ranging from $_{30}Zn^{72}$ up to $_{66}Dy^{161}$ have been measured for both uranium and plutonium fissions. For nondestructive burnup determination the γ-emitting fission isotopes may be used, since the best suited definition of burnup is fission density.[1] This is the number of fissions which have occurred per unit volume of fuel material. In most cases burnup values corresponding to other definitions, e.g., percent depletion of fissioned isotopes, or total energy produced, or yield, can be inferred directly from the basic quantity fission density. A record of the fission density exists in the fuel itself, namely in the form of the fission products. Some of them can be determined nondestructively by γ spectroscopy.

If 1 g of U^{235} is fissioned in a burst, an activity of approximately 10,000 C is obtained. Approximately 100 days of cooling reduces the activity by a factor of 100. Show that this leads to an activity of about 3×10^{12} disintegrations per second. According to Wigner and Way[2] the ratio of β particles to γ quanta is roughly 2 : 1. We therefore have available

[1] R. L. Stower, and G. K. Moeller, Mass. Inst. Technol. Rept. MIT-OR 6, 1961.

[2] K. Way, and E. P. Wigner, *Phys. Rev*. **73,** 1318 (1948).

* Problem 31 is by M. J. Higatsberger.

1×10^{12} γ quanta every second, three months after 1 g of U^{235} is completely fissioned.

If we proposed to measure the γ-ray spectrum of the fission products with the conventional NaI crystal attached to a photomultiplier, a preamplifier, and a multichannel analyzer, we would find that resolution, background, and the strong energy dependence of efficiency in the NaI crystal does not permit solving the problem. In recent years, however, semiconductor detectors and improved electronic circuits have become attractive tools in γ spectroscopy. Originally lithium-drifted silicon semiconductors and later lithium-drifted germanium detectors have become available. With such detectors a resolution of at least 10 keV fwha is obtainable; however, by extending the measurements from higher to lower energies heavy background is found in which a number of characteristic γ lines are completely hidden, partly due to the Compton continuum of higher energy γ lines.

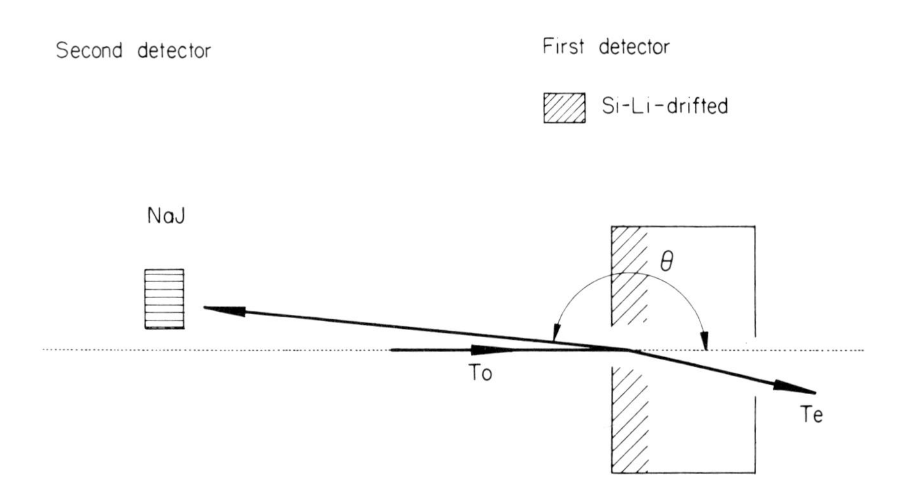

FIG. 31.1. Principle of Compton mode operation.

31.2.2. Principle of the Semiconductor Compton Spectrometer

In Fig. 31.1 a γ quantum with the energy T_0 is shown impinging on a semiconductor detector with a depletion layer up to 10 mm for silicon and

5 mm in case of germanium. The Compton electron is produced with an energy

$$T_e = \frac{T_0}{1 + m_e c^2/[T_0(1 - \cos\theta)]}. \tag{31.1}$$

The rest of the energy T_S is carried off by the scattered photon and depends on the angle θ. The term T_0 is the incident γ-ray energy and $m_e c^2$ is the electron rest mass. The angle θ is defined by the direction of the incoming and scattered photons. Using conservation of energy, express the energy T_0 in terms of the energies T_e and T_S.

Compton-mode operation is achieved by locating the detectors as shown, measuring the scattered γ rays in the second detector, and the Compton electrons in the first. With proper electronic circuits time-coincident events only are recorded. By differentiation of Eq. (31.1), show that the Compton electron energy is at a maximum for $\theta = 180°$. What is the significance of the extremum for $\theta = 0°$? What is the Compton electron energy for $\theta = 0°$? Solve Eq. (31.1) for the original photon energy T_0. Evaluate this result for $\theta = \pi$ and graphically express T_0 as a function of T_e for $0 \leqslant T_e \leqslant 2$ MeV.

In order to build a suitable experimental apparatus, means must be provided for the collimation of the primary beam of γ rays, which must be directed towards a semiconductor with an appropriate depletion layer; and the back-scattered quanta must be recorded approximately 180° backwards by a second detector.

31.2.3. Prototype Spectrometer

Constructing a prototype instrument could proceed as follows (see Fig. 31.2):

From the fuel element ① γ radiation passes through collimators ②, ④, and ⑤ to the semiconductor ⑥, after having passed a throughhole of the NaI crystal ⑦; ⑧ is the photomultiplier of the NaI crystal and ⑬ the preamplifier belonging to the same unit. Further important elements of the instrument are the vacuum mounting of the semiconductor ⑨, the cooling device ⑩ (either liquid air, dry ice, or Peltier cooling), and the preamplifier for the semiconductor ⑭; ⑫ is shielding, and ⑮ indicates the mounting base and ⑯ the mounting frame. The connection ⑪ leads to a vacuum pump. In practical cases this connection is sealed off. Known absorbers ③ may be inserted and withdrawn to allow determination of the self-absorption coefficient when tested with known radiation. Preferably collimators ②, ④, and ⑤ should be adjustable, in order to give the instrument a wide operating range to cover fuel with considerably

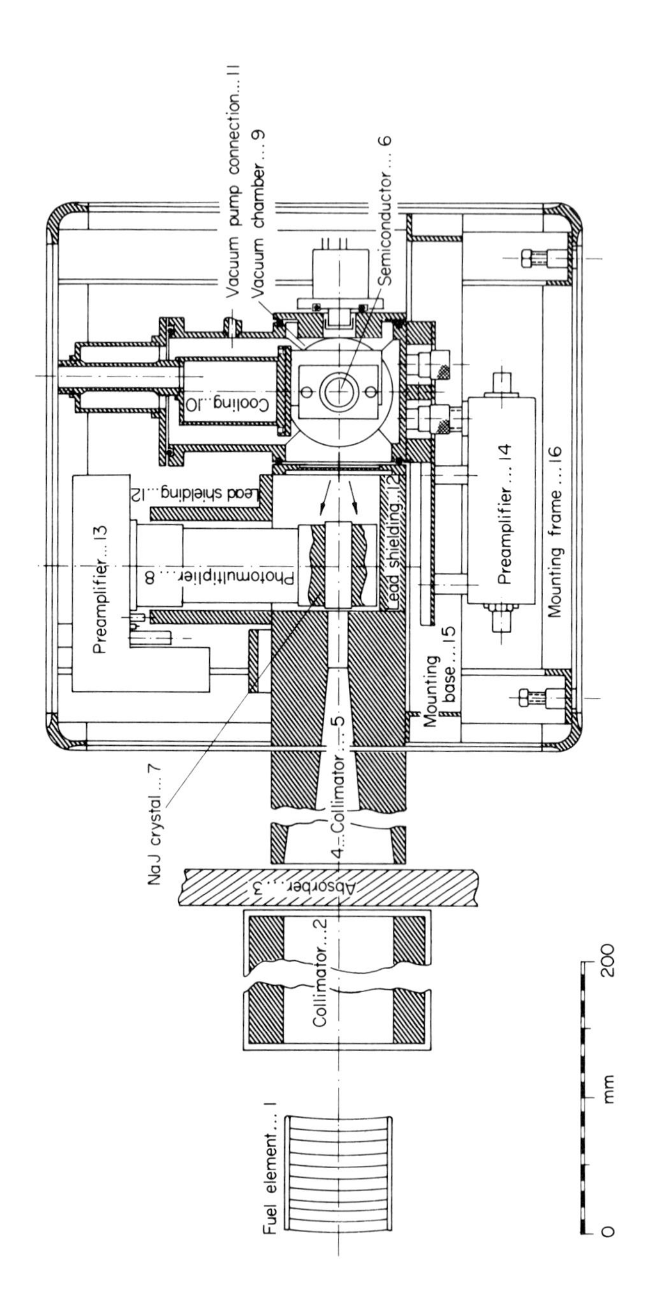

Fig. 31.2. Prototype Compton coincidence spectrometer.

varying radioactive source strength. If only a portion of the fuel element is seen through a collimator, scanning is required along the fuel element for burnup measurements of the total element.

The spectrometer head including the heavy metal collimator may weigh less than 100 lb, with overall dimensions 40 × 40 × 20 cm. Optimum performance will be achieved by carefully arranging the collimator to prevent scattered γ rays from reaching the semiconductor. A NaI crystal is chosen as second detector with a crystal size of 2 × 2 in.; a hole of $\approx$ 20-mm diameter is necessary to avoid interaction with the primary γ-ray beam. Aluminum lining will improve guidance of the light onto the photomultiplier.

31.2.4. Operation

Satisfactory operating performance of the instrument will depend on stable, low noice electronics, good geometry of the collimators, and good resolution of the primary detector. Examine the current literature to determine the best resolution achievable with a lithium-drifted, liquid-air cooled silicon detector.

The count rate must be limited to about 5000 cps. The maximum permissible number of γ quanta reaching the semiconductor per unit time has to be adjusted by distance and collimator geometry. If a specific fuel activity corresponding to 300 mg of U^{235} fissioned/cm length of fuel element is assumed, a reduction factor of approximately 2×10^{-9} is required, if a 1-cm length of an MTR fuel element is to be seen by the detector. An amount of 300 mg of U^{235} fissioned corresponds, after some cooling, to roughly 50–80 Ci total γ activity and approximately 1% thereof is attributed to Cs^{137}. If the open area in front of the detector is approximately 10 mm^2, the distance of the fuel element under investigation and the collimator slit width must be adjusted accordingly. Show that the following simple formula holds:

$$\frac{G_{U^{235}} R S}{L^2} = 10^4, \tag{31.2}$$

where L is the distance from the fuel element to the spectrometer head in meters, S is the slit width of the collimator in millimeters, and R is the activity ratio (total activity to activity of isotope under investigation) for $G_{U^{235}}$ mg of U^{235} fissioned/cm fuel element length.

31.2.5. Determination of Burnup History

Radiochemical separations were until recently the only means of reasonable fission product analysis. In a few cases also complicated mathematical analyses of γ-ray spectra were successful. Since now also high resolution γ spectroscopy exists, it is possible to measure the characteristic isotope directly. The most suitable isotope, because of its very long half-life, is Cs^{137} with a half-life of 29.8 years. This fission product has a cumulative yield of 6.15%. By what decay mode does this isotope produce a single, strong, γ ray of 0.662 MeV? Determining the intensity of Cs^{137} is a measure of all fissions which have occurred. Other suitable isotopes are given in Table I. In Fig. 31.3 the γ activity of an MTR element is

TABLE I. Important γ-Active Fission Products for Burnup and Burnup History Determination

			Main γ lines	
Isotope	$T/2$	Cumulative yield (%)	(MeV)	Intensity in percentage per decay
Ru^{103}	39.7 *d*	3.0	0.610	6
			0.555	0.5
			0.495	90
			0.440	0.5
			0.055	1.0
Ba^{140}	12.8 *d*	6.35	0.537	30
			0.436	5
			0.304	5
			0.162	7
La^{140}	40.2 *h*	6.32	2.523	5.5
			1.597	94
			0.926	11
			0.816	35
I^{131}	8.14 *d*	3.1	0.724	2.8
			0.638	9.3
			0.364	80.9
			0.284	6.3
Zr^{95}	65 *d*	6.2	0.756	43
			0.724	55
			0.235	2
Nb^{95}	35 *d*	6.1	0.768	99
Ce^{144}	284.5 *d*	6.0	0.134	17.1
Pr^{144}	17.5 *m*	6.0	0.695	1.5
Ru^{106} (Rh^{106})	1.0 *a*	(30 *s*) 0.38	0.624	9.9
			0.513	20.5
Cs^{137} (Ba^{137})	29.8 *a*	(2.6 *m*) 6.15	0.662	92

shown as a function of γ-ray energy, with the peaks attributable to various fission products labelled with isotope, half-life, and γ-ray energy. The time dependence of the γ yield for some fission products related to total

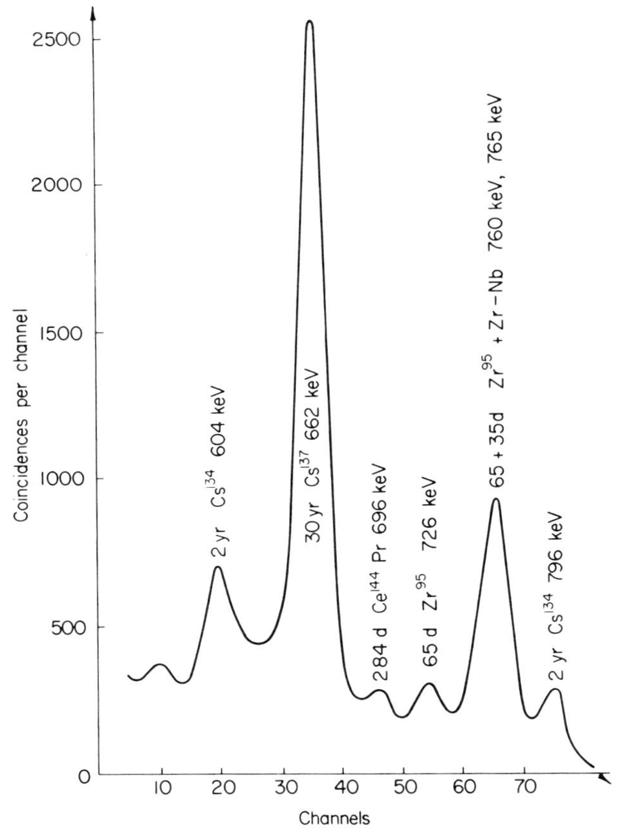

FIG. 31.3. Sum spectrum of MTR fuel plate (total of 18 pieces).

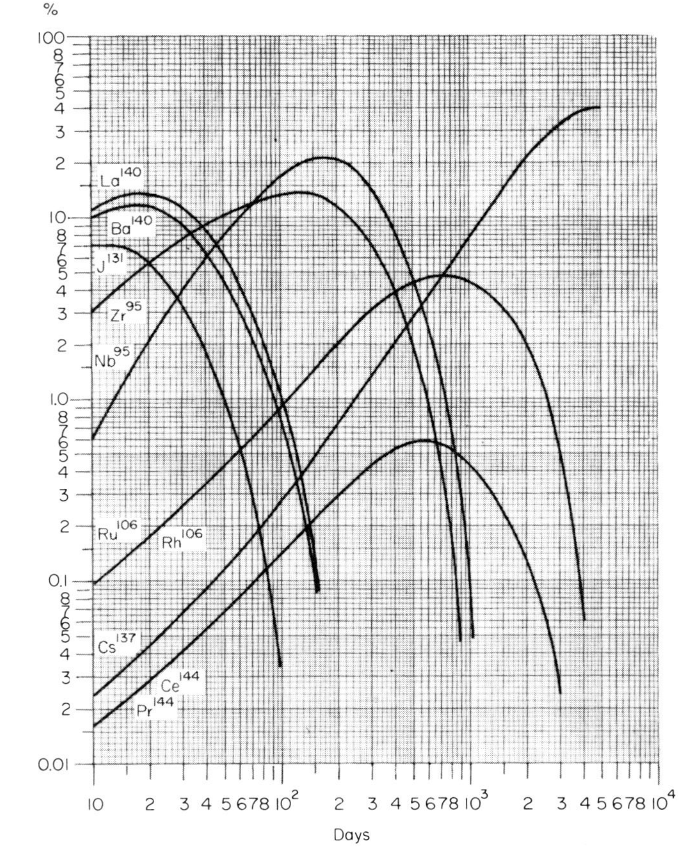

FIG. 31.4. Time dependence of γ yield for some fission products related to total γ activity produced by a neutron burst.

γ activity is shown in Fig. 31.4. The assumption here is that the fission products were produced in a burst. When the reactor operating history is given with time-dependent neutron fluxes, only a three-dimensional representation reveals the necessary information. From Fig. 31.4 and the values listed in Table I, one can predetermine by measuring the relative intensity of at least two fission products the cooling time involved. For a more correct mathematical treatment the number of fission isotopes produced is determined by:

$$\frac{dN_2}{dt} = -\lambda_2 N_2 - \sigma_2 \Phi N_2 + \lambda_1 N_1 + \gamma_2 \Sigma_f \Phi, \tag{31.3}$$

where $\lambda_2 N_2$ is the radioactive decay rate and $\sigma_2 \Phi N_2$ the neutron capture rate of N_2, while N_2 is formed at the rate $\lambda_1 N_1$ by the decay of its precursor; there is also a direct production term of N_2 given by fission yield, fission cross section, and flux $\gamma_2 \Sigma_f \Phi$. Show that the solution of this differential equation is (hint: use the integrating factor $\exp[(\lambda_2 + \sigma_2 \Phi)t)]$

$$N_2(t) = \frac{\lambda_1}{\lambda_1 + \sigma_1 \Phi} \gamma_1 \Sigma_f \Phi \left[\frac{1 - \exp[-(\lambda_2 + \sigma_2 \Phi)t]}{\lambda_2 + \sigma_2 \Phi} + \frac{\exp[-(\lambda_2 + \sigma_2 \Phi)t] - \exp[-(\lambda_1 + \sigma_1 \Phi)t]}{\lambda_2 + \sigma_2 \Phi - \lambda_1 - \sigma_1 \Phi} \right] + \frac{\gamma_2 \Sigma_f \Phi}{\lambda_2 + \sigma_2 \Phi} (1 - \exp[-(\lambda_2 + \sigma_2 \Phi)t]. \tag{31.4}$$

In case N_1 has a very short half-life ($\lambda_1 \approx \lambda_1 + \sigma_1 \Phi$), what is the result for $N_2(t)$?

Another interesting case is $\lambda_2 = 0$ (i.e., N_2 is stable). What is the result for $N_2(t)$ in this case?

Assuming that after $N_2(t)$ was formed in equilibrium ($dN_2/dt = 0$) at flux Φ, Φ is reduced to Φ', show that the new resulting value N_2' will be

$$N_2'(t) = \left(\frac{\lambda_1}{\lambda_2 - \lambda_1}\right) \frac{\lambda_1 \Sigma_f \Phi'}{\lambda_1 + \sigma_1 \Phi'} [\exp(-\lambda_1 t) - \exp(-\lambda_2 t)] + \left[\frac{\lambda_1 \gamma_1 \Sigma_f \Phi'}{(\lambda_1 + \sigma_1 \phi')(\lambda_2 + \sigma_2 \phi')} + \frac{\gamma_2 \Sigma_f \Phi'}{\lambda_2 + \sigma_2 \Phi'}\right] \exp(-\lambda_2 t) \tag{31.5}$$

with t here the time after flux Φ has changed to Φ'.

31.3. Exercises

31.3.1.

Compare a conventional Compton spectrometer with a semiconductor-type instrument.

31.3.2.

Assume that for the Compton spectrometer both primary and secondary detectors are to be semiconductors. What advantages and disadvantages result?

31.3.3.

What would an instrument as described under Eq. (31.3) look like when portability and battery operation are required?

31.3.4.

Draw time dependence with respect to the yield for the fission products of Table I for several different (constant) flux levels.

32. SPECTROMETER COMPARISON*

32.1. Problem

Make a comparative study of the design considerations for photo-peak spectrometers, anti-Compton spectrometers and Compton spectrometers, using equal size Ge-Li-drifted primary detectors. Show the basic equipment involved for all these spectrometers and determine the response functions and efficiencies.

32.2. Solution

32.2.1. Background

Nuclear particle research has been greatly enhanced since semi-conductor detectors have been in use, but in γ spectroscopy in particular entirely new possibilities were found. Scintillation counters were a big stride forward in γ spectroscopy, but energy resolutions better than approximately 7% were never achieved.

Semiconductor detectors showed almost immediately at their introduction that resolutions could be increased by an order of magnitude or better. Energy resolutions of 0.10% or better are now more the rule than the exception. The efficiency was also improved in the region of high γ energies by employing new large volume Li-drifted semiconductors. Going hand in hand with better and bigger Si or Ge detectors are improvements related to low noise electronic instrumentation.

For γ spectroscopy three basic types of instruments may be envisaged, namely the full energy peak spectrometer, the Compton spectrometer, and the anti-Compton spectrometer. To facilitate the choice of the most suitable instrument for a given application we shall compare the three spectrometer types for the same size semiconductor detector and similar instrumentation. Of interest in the comparison are the response functions, the intrinsic peak efficiencies, and the resolutions. As the amount of instrumentation necessary also varies among the instruments, the economic aspect is to be considered also.

* Problem 32 is by M. J. Higatsberger.

A comparative study of the three spectrometers in question for the special case of scanning irradiated fuel samples has been undertaken by Hick.[1]

32.2.2. Photo-Peak Spectrometer

The photo-peak spectrometer uses the full energy peaks of γ rays for spectrum analysis. Ge(-Li) detectors are used in a photo-peak spectrometer rather than Si(-Li) detectors. Why?

The spectrometer consists of a Ge(-Li) detector, mounted inside a vacuum cryostat, operated at liquid nitrogen temperature. Why must the detector be operated at liquid nitrogen temperature? In order not to cause peak tails or low energetic background due to scattering the collimated primary beam should pass only thin windows as well as the semiconductor detector. List the types of electronic units needed to operate a photo-peak spectrometer and estimate their cost on the basis of current prices.

32.2.2.1. Intrinsic Efficiency and Response Function. For a given source, the total count rate is the same for all three spectrometer types, because they have been chosen to use the same collimator geometry and the same detector, a 20-mm diameter × 5-mm Ge(-Li) detector with the γ beam parallel to the axis of rotation. This count rate is limited to 10^3–10^4 Hz because of pulse pileup and can be determined by use of the total intrinsic efficiency values shown in Table I. For a strong source the

TABLE I. RESPONSE OF A Ge(-Li) DETECTOR, 20-MM DIAMETER × 5 MM (GEOMETRY FIGS. 32.2., 32.3)

(1) E (keV)	(2) Total intrinsic efficiency (%)	(3) Simple theoretical photo efficiency (%)	(4) Experimental full energy efficiency (%)	(5) Simple theoretical Compton (%)	(6) Experimental Compton (%)	(7) Experimental DEPP (%)	(8) Experimental SEPP (%)
100	74	55	63	19	11	—	—
200	35.4	9.4	14	26	21.6	—	—
400	21.3	0.6	3.1	20.7	18.2	—	—
800	15.2	0.2	0.8	15	14.4	—	—
1200	12.6	~ 0	~ 0.35	12.5	12.2	~ 0.02	~ 0
1600	10.9	—	~ 0.25	10.85	10.5	~ 0.11	~ 0.003
2400	9.2	—	(~ 0.15)	8.75	8.7	~ 0.33	(~ 0.01)
3200	8.6	—	(~ 0.05)	7.5	8	~ 0.57	(~ 0.02)

[1] H. Hick, *Nucl. Instr. Methods* **40,** 337 (1966).

tolerable pulse pileup determines the solid angle of the collimator via the total intrinsic efficiency. For very weak sources the maximum possible solid angle of an anti-Compton or Compton arrangement is limited because of geometric considerations imposed by the method of operation.

The total intrinsic efficiency is the fraction of primary γ quanta which have transferred some energy to the detector and counted due to the charge collected.

It is given by

$$I_{tot} = 1 - e^{-\tau x}, \tag{32.1}$$

where $\tau = (u + \sigma + K)$ is the total linear attenuation coefficient; u is the photoelectric linear absorption coefficient; σ is the Compton total linear attenuation coefficient; K is the pair production linear attenuation coefficient; and $x = 0.5$ cm is the Ge(-Li) detector thickness. The calculation of I_{tot} is based on cross section data[2] for Ge and the results are shown in Table I, column 2.

For a photo-peak spectrometer the useful part of the spectrum is the area of the full energy peak which has a nearly Gaussian shape. If one neglects multiple events, a straightforward calculation with the formula

$$I_{photo} = (u/\tau)(1 - e^{-\tau x}) \tag{32.2}$$

yields the intrinsic photo-peak efficiency I_{photo} shown in Table I, column 3. An accurate calculation of the response function would require a very complicated Monte-Carlo calculation for photon and electron histories in order to take into account multiple events and electron range effects. Therefore, an extrapolation of available experimental data[3] has been used to obtain the actual full energy peak intrinsic efficiency, shown in Table I, column 4.

The intrinsic Compton continuum efficiency $I_{Compton}$ may be calculated simply as

$$I_{Compton} = (\sigma/\tau)(1 - e^{-\tau x}), \tag{32.3}$$

and is shown in column 5 of Table I.

The intrinsic double escape pair peak efficiency I_{pair} would be given by

$$I_{pair} = (K/\tau)(1 - e^{-\tau x}). \tag{32.4}$$

For comparison, experimental data have been extrapolated to obtain the single (SEPP) and double escape pair peak (DEPP) intrinsic efficiencies shown in columns 7 and 8.

[2] G. W. Grodstein, NBS Circular 583 (1957).

[3] G. T. Ewan and A. J. Tavendale, *Can. J. Phys.* **42**, 11 (1964).

The experimental Compton continuum intrinsic efficiency is obtained as the difference of columns 7, 8, and 4 to column 2 and is shown in column 6. It also contains a small contribution of peak tails and background.

32.2.2.1.1. Questions.

1. Why are experimental full energy peak fractions always higher than theoretical values, especially at high energies?

2. How do multiple events and electron range effects account for the fact that the experimental Compton continuum fraction is smaller than the theoretical value up to 2.4 MeV, and the experimental value is higher than the simple theoretical value for energies higher than 2.4 MeV?

3. Why are the experimental pair peaks smaller than theoretically predicted? These deviations of actual values from simple theoretical ones depend strongly on the size and the geometry of the detector and the primary beam direction relative to the detector.

Based on the experimental values of Table I it is possible to establish the response functions for various energies shown in Fig. 32.1. The response functions are shown as they would be obtained using a 256-channel analyzer, with the full energy peak at channel 240 and for 10^6 incident quanta. Peak shapes have been assumed to be Gaussian, which is a good approximation. The full width at half-maximum (fwhm) has been assumed to rise linearly from 5 keV at 100 keV, to 10 keV at 3.2 MeV, which seems reasonable in routine operation. Small peak tails are shown, based on typical experimental values. The amount of these peak tails depends on the quality of the semiconductor detector, its operating conditions, scattering in source, collimator, and cryostat. These peak tails are present in comparable amounts in the spectra of all three spectrometer types. But while they are almost negligible—in comparison with other contributions—for the photo-peak and anti-Compton spectrometers, they are a major fraction of the peak tails of the Compton spectrometer. For the Compton continuum a shape has been assumed which is predicted theoretically for single events by the differential cross section $d\sigma/dT$ for giving a free electron a recoil energy within the interval $(T, T + dT)$. A correction for multiple events has been applied in drawing experimentally observed double Compton steps.

Two further differences exist from actually observed spectra.

a. The peaked region near the Compton steps is in practice flattened due to multiple events.

b. The low energy region is in practice increased due to a low energy background of multiply scattered γ rays. This effect does not belong to the response function; but because of lack of suitable approximation methods these two effects have been ignored.

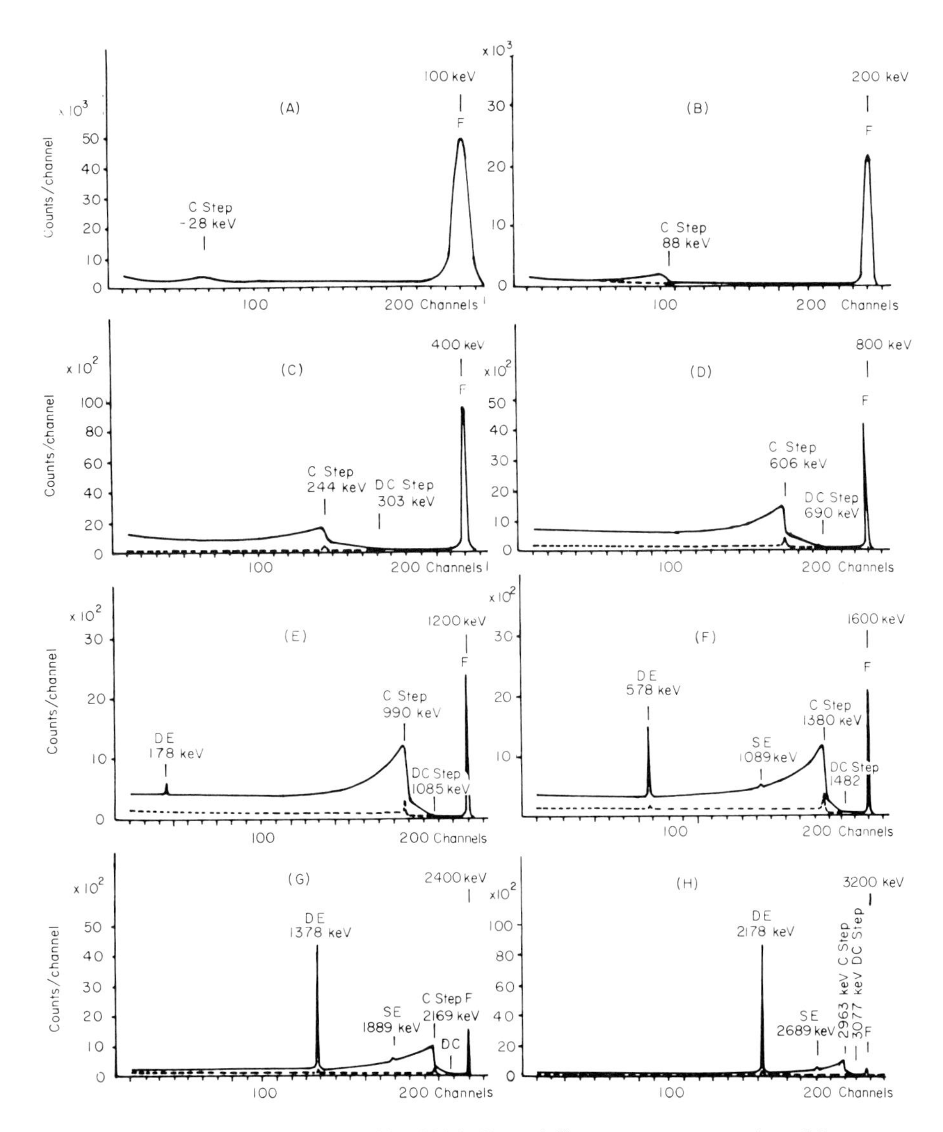

FIG. 32.1. Direct spectrum, 100–3200 keV; anti-Compton spectrum, dotted lines.

32.2.2.2. Discussion of Performances. At low energies (up to ~ 400 keV) what can be said about the usefulness of the response function?

For energies up to ~ 1200 keV how is the response function modified? What property of the full energy peak efficiency facilitates spectrum analysis of a multi-γ-ray spectrum for energies up to ~ 1200 keV. What difficulties may be encountered in the quantitative determination of peak areas in such a spectrum?

For higher energies than 1200 keV what complications arise in the response function? How could these complications effect the analysis of a multi-γ-ray spectrum? What peak would be most useful for making accurate γ-ray energy determinations at energies higher than 1200 keV?

32.2.3. Anti-Compton Spectrometer

The anti-Compton spectrometer, like the photo-peak spectrometer, uses the full energy peak for spectrum analysis. One tries to suppress the interfering Compton continuum by surrounding, as completely as possible, the Ge(-Li) detector with a large anticoincidence NaI crystal. A Compton event yielding a detector pulse within the Compton continuum should be accompanied by the detection of the scattered quantum in the NaI crystal. By use of an anticoincidence such events should be eliminated.

List the types of electronic equipment needed to operate an anti-Compton spectrometer and estimate their cost on the basis of current prices.

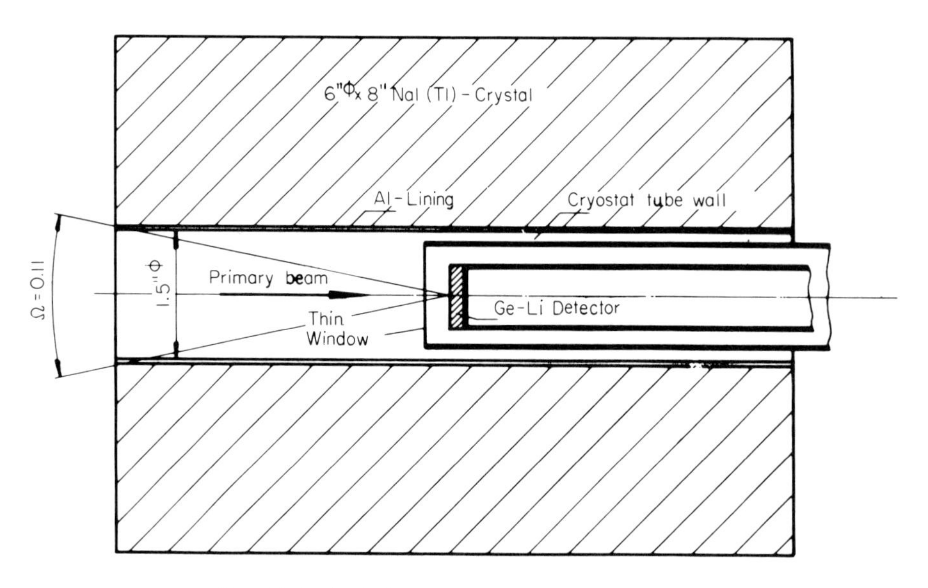

FIG. 32.2. Anti-Compton spectrometer geometry.

32.2.3.1. Intrinsic Efficiency and Response Function. Apart from the danger of chance anticoincidences, which would remove full energy events, the intrinsic full energy peak efficiency is the same as for the photo-peak spectrometer.

A possible geometry is shown in Fig. 32.2. A 15 cm diameter × 20 cm NaI (T1) crystal with a 3.8 cm diameter throughhole surrounds the semiconductor detector, which is mounted on a cold finger. The "dead angle" Ω should be as small as possible. Note that there are two dead angles, one in the direction to, and one in the direction from, the source. This "dead angle" causes a peak at the Compton step. Why? The efficiency for this peak can be calculated in much the same way as the efficiency for the Compton spectrometer to be discussed later. For our geometry this efficiency is two-thirds of that for the Compton spectrometer L_1 (Subsection 32.2.4.3).

The "removal efficiency" for the rest of the Compton continuum depends on the intrinsic efficiency of the NaI crystal (above the discriminator level) and the absorption in the cryostat tube wall and the aluminum lining of the throughhole of the scattered γ rays. Assuming a typical value of 2-mm aluminum thickness the fractions shown in Table II will escape the anticoincidence detection.

TABLE II.

Energy of back-scattered quantum (keV)	Fraction escaping (%)
100	~ 12
200	8
400	13
800	21
1200	33
1600	37
2400	42
3200	44

Why does this fraction increase at very low energies where the intrinsic efficiency of the NaI crystal is nearly 100%? Why does this fraction increase for high energies? The "dead solid angle" causes spurious single- and double-escape pair-peaks. Why?

Another cause for spurious pair-peaks is the absorption loss and the low intrinsic efficiency of the NaI crystal for 511-keV quanta. Show that for the single-escape pair peak:

1. A fraction of $2\Omega(4\pi) = 1.8\%$ escapes through the dead solid angles (2Ω).
2. A fraction of 4.5% is stopped before reaching the NaI crystal.
3. A fraction of 10% passes the crystal without being detected.

Thus, the single-escape pair-peak is reduced to 16% of its original value. For the double-escape pair-peak, two quanta have to escape detection simultaneously in order for that event not to be detected by the anti-coincidence system. Using the information given for the single-escape pair-peak, show that the double-escape pair-peak is reduced to 3% of its original value.

By means of these reduction values it is possible to obtain the response function of the anti-Compton spectrometer based on the response function of the photo-peak spectrometer. The curves are shown with dotted lines in the graphs of the photo-peak spectrometer (Fig. 32.1).

32.2.3.2. Discussion of Performances. How does the response function compare to that for the photo-peak spectrometer for low energies (up to $\sim$ 400 keV)? for higher energies? Are there still complications in the response function which could effect the analysis of a multi-γ-ray spectrum? What is the relative importance of these complications in the anti-Compton and photo-peak spectrometers? How would the performance and response function of the anti-Compton spectrometer be effected if the big anti-coincidence crystal were not shielded against background? How would the response function be effected if the primary beam were allowed to hit any material other than the Ge(-Li) detector (e.g., the NaI crystal)?

Indicate how the efficiency calibration would have to be modified to study γ rays in cascades.

32.2.4. Compton Spectrometer

A suitable geometry for a Compton spectrometer is shown in Fig. 32.3. A primary γ beam passes a 5 $\times$ 5 cm NaI crystal, without any interactions, through a cylindrical hole, and hits a Ge(-Li) detector. For energies below 1.02 MeV only Compton scattering events cause secondary γ quanta of energy T_S to leave the semiconductor detector in various directions.

List the types of electronic equipment needed in addition to the photo-peak spectrometer equipment to operate a Compton spectrometer and estimate their cost on the basis of current prices.

32.2.4.1. Resolution. There are two further contributions to the line width in a Compton spectrometer:

1. a geometric contribution due to the range of back-scattering angle accepted by the NaI crystal; and

2. the fact that the electrons are not really free and at rest, as supposed in the Compton formula, but have a momentum distribution.

If Compton-scattered secondary γ rays are observed by means of a NaI

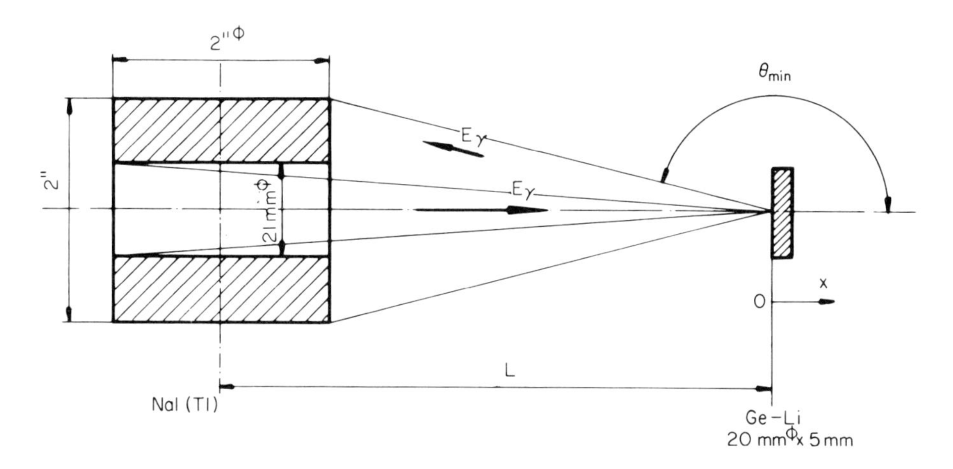

FIG. 32.3. Compton spectrometer geometry: $L_1 = 125$ mm, $L_2 = 75$ mm, $L_3 = 50$ mm; $\theta_{\min} = 166°$, $\theta_{\min} = 154°$; $\theta_{\min} = 135°$; $\theta_{\max} = 176°$, $\theta_{\max} = 174°$, $\theta_{\max} = 172°$.

crystal for a scattering angle range $\Delta\theta$ the corresponding Compton electron energy varies according to:

$$\Delta T_e = (\delta T_e/\delta\theta)_\theta \, \Delta\theta. \tag{32.5}$$

To observe a Compton peak of best resolution ΔT_e has to be minimized. Show that $(\delta T_e/\delta\theta)_\theta$ is a minimum at $\theta = 180°$, and that the resolution is best for a back-scattering geometry with $\theta \sim 180°$ and $\Delta\theta$ as small as possible. What is the value of $(\delta T_e/\delta\theta)_\theta$ at $\theta = 180°$?

For three values of L (Fig. 32.3) the amount of ΔT_e (corresponding to $\Delta\theta = \theta_{\max} - \theta_{\min}$, Fig. 32.3) is shown in Table III.

TABLE III.

T_0 (keV)	ΔT_e (keV)		
	L_1	L_2	L_3
200	1	3	7.5
1200	2	9	28
3200	4.5	15	35.5

The effect of bound electrons on the line width increases with the atomic number of the material. Why? A Ge detector ($Z = 32$) is less favorable in this respect than a Si detector ($Z = 14$). Also, with respect to spurious pair-peaks, a Si detector would be preferable for a Compton spectrometer. Why? An advantage of the Ge detector is its higher efficiency and better energy-to-charge conversion, which improves resolution.

This effect of bound electrons is mainly of importance for low energies (being negligible for energies greater than 400 keV). Is this observation reasonable?

32.2.4.2. Nonlinearity of Response. Show that the Compton electron energy T_e converted to a charge pulse, for a Compton-scattered γ energy T_S smaller than the primary γ energy T_0, can be expressed by

$$T_e = T_0[1 - \{1 + \{T_S/(m_0c^2)\}(1 - \cos\theta)\}^{-1}]. \tag{32.6}$$

According to this relation the Compton electron spectrum is slightly nonlinear. For a first approximation ($\theta = 180°$) show that

$$T_e = T_0 - 200 \text{ keV}, \tag{32.7}$$

which relates the Compton peaks (corresponding to T_e) with the full energy T_0.

32.2.4.3. Efficiency. The number of quanta scattered from the volume element dV of the detector into the solid angle element $d\Omega$ is given by

$$\frac{\partial}{\partial\Omega}\frac{(\partial J')}{\partial x}\, dV\, d\Omega = J_0 nZ \frac{d\sigma}{d\Omega} \exp\{-(\Sigma + \Sigma')x\}\, dV\, d\Omega, \tag{32.8}$$

where J_0 is the incident number of gamma quanta/cm^2 sec; J' is the back-scattered number of gamma quanta/cm^2 sec; x is the coordinate of volume element dV, parallel to the primary beam, for the front face of the detector at $x = 0$; nZ is the electron density for Ge $\sim 14.1 \times 10^{23}\ e^-/\text{cm}^3$ and for Si $\sim 7 \times 10^{23}\ e^-/\text{cm}^3$; $d\sigma/d\Omega$ is the differential (Klein–Nishina) cross section per electron (cm^2/sra); Σ is the total macroscopic cross section of detector material at energy T; Σ' is the total macroscopic cross section of detector material at energy T'; dV is the volume of the detector, $dV = dF\, dx$; and $d\Omega$ is the solid angle element.

Show that Eq. (32.8) can be written as

$$J' = J_0 nZ \int (d\sigma/d\Omega) d\Omega \int \exp\{-(\Sigma + \Sigma')x\}\, dx. \tag{32.9}$$

The integration over $d\Omega$ has to be done for the Ω for which the NaI crystal is seen by the semiconductor detector. The integration over dx has to be done over the sensitive detector thickness ($x = 0.5$ cm).

Data for Σ and Σ';[2] data for $d\sigma/d\Omega$.[4] The integration of $(d\sigma/d\Omega)\, d\Omega$ in fact gives slightly different results for the three geometrical cases L_1, L_2,

[4] A. T. Nelms, NBS Circular 542 (1953).

L_3, but these differences are so small that they have been neglected. The results of the integrations are given in Table IV where $A = \int (d\sigma/d\Omega)\, d\Omega$ and $B = \int \exp\{-(\Sigma + \Sigma')x\}\, dx$.

The number of back-scattered quanta, reaching the 2 × 2 in NaI crystal has to be multiplied by the detection efficiency of the crystal (column D of Table IV). In this detection efficiency the absorption losses of typically 1-mm aluminum between the Ge detector and NaI crystal are included, but the finite solid angle of the NaI crystal is not included. Determine the solid angle Ω_1, Ω_2, Ω_3, for geometries L_1, L_2, L_3 and show that the resultant intrinsic efficiencies are given by I_1, I_2, I_3 in Table IV.

TABLE IV.

E (keV)	A ($\times 10^{-26}$)	B ($\times 10^{-2}$)	D	I_1 ($\times 10^{-4}$)	I_2 ($\times 10^{-4}$)	I_3 ($\times 10^{-3}$)
100	4.20	9.7	0.9	9.74	14.5	9.2
200	2.85	23.6	0.88	15.25	22.9	14.4
400	1.72	34.7	0.85	13.33	20	12.6
800	1	37.9	0.825	7.93	11.9	7.48
1200	0.72	38.6	0.825	5.8	8.7	5.48
1600	0.59	39	0.285	4.71	7.06	4.45
2400	0.39	39.4	0.825	3.09	4.64	2.92
3200	0.3	39.7	0.825	2.42	3.63	2.28

32.2.4.4. Spurious Peaks. Spurious peaks are possibly due to chance coincidences and to pair production. The effect of chance coincidences is negligible if the shielding of the NaI crystal against background is good enough. Why? The time-correlated coincidences between a pair production event in the semiconductor detector and an annihilation quantum detected in the NaI crystal may be minimized by a suitable discrimination of NaI pulses. Show that the Compton scattered quanta have an energy range from 72 to 237 keV. In this energy range almost all of them will be recorded in the photo peak of the NaI spectrum. The annihilation quanta of 511 keV will fall into this region to a fraction of 25% at a total intrinsic efficiency of ~ 70%. Therefore only a fraction of $2\{\Omega/(4\pi)\}\ 0.25 \times 0.70$ of the double escape pair peaks observed with the photo-peak spectrometer will be seen in the Compton spectrum. What is this fraction for L_1, L_2, and L_3? What are the single-escape pair-peak fractions? What other steps could be taken to reduce chance coincidences?

32.2.4.5. Response Functions. Based on the data on resolution, efficiency, and spurious pair-peak fractions, response functions have been established (Fig. 32.4). The peak tails observed in practice are *mainly* due to two

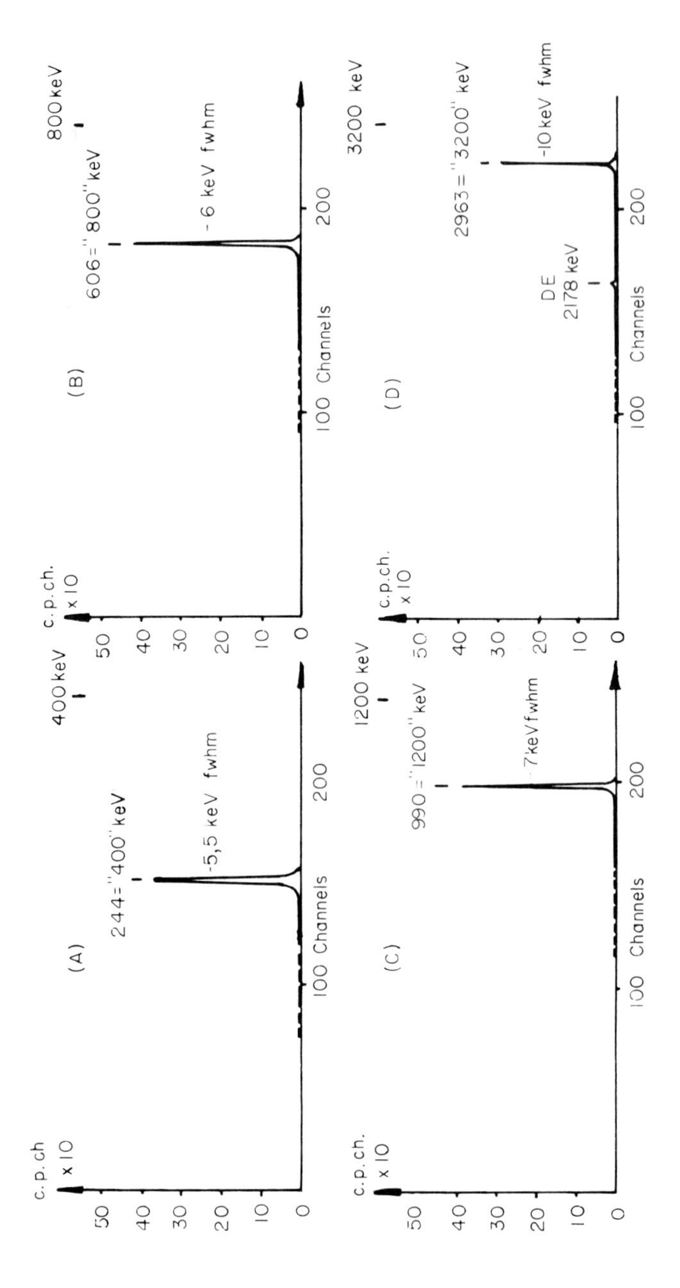

FIG. 32.4. Coincident spectrum 400–3200 keV.

successive Compton-scattering events. Explain how such events can produce peak tails such as are observed in Fig. 32.4. Does the peak tail reduce the calculated peak efficiency? Further small peak-tail contributions are due to scattering in source and collimator and do not belong to the spectrometer response function. Can you suggest any peak-tail contributions which are due to characteristics of the Ge(-Li) detector itself (and therefore not characteristic of the Compton operation mode)? The peak tails shown in Fig. 32.4 are typical experimental values of peak tails obtained without special precautions to eliminate double-scattering events. How does this response function compare with those of the spectrometers previously discussed?

32.2.4.6. Discussion of Performance. Compare the efficiency and resolution of the Compton spectrometer and the photo-peak spectrometer at low energies. What difficulties might be encountered with the Compton spectrometer at low energy, but would not be a problem with the photo-peak spectrometer?

For energies higher than 400 keV, how does the energy resolution and efficiency of the Compton spectrometer compare with the photo-peak spectrometer?

32.2.5. Comparison of Spectrometers

32.2.5.1. Efficiency. At low energies how does the efficiency of the photo-peak and anti-Compton spectrometers compare with the Compton spectrometer? At energies $\sim$ 400 keV what must be sacrificed with a Compton spectrometer to achieve an efficiency comparable to an anti-Compton spectrometer?

At higher energies which spectrometer has the smallest efficiency? Approximately how much smaller is it?

32.2.5.2. Peak Fractions. The peak fraction is a figure that reflects the quality of the response function. Which spectrometer has the smallest peak fraction at all energies discussed? Which has the largest? Using the calculated response functions, make a rough graph of the energy dependence of the peak fractions from 400 to 3200 keV.

The useful efficiency to determine the intensity of a single peak is the intrinsic peak efficiency. Which spectrometers have large peak efficiencies? However, the useful efficiency to analyze a complex spectrum is some combination of intrinsic peak efficiency and peak fraction, for instance, the product, because the reciprocal peak fraction is a measure of the amount of background produced by a peak. Compare this product for the three spectrometers at high energy ($\sim$ 1 MeV) and at low energy ($\sim$ 400 keV).

Use the price estimates to determine the relative cost of the three spectrometers discussed. Which is the most expensive? If you wanted to but a γ-ray spectrometer on a limited budget, what basic equipment would you buy and why?

32.3. Exercise

What is the direction of the ejected electron in the photoelectric, the Compton, and the Rayleigh scattering processes?

33. PAIR PRODUCTION CROSS SECTION*

33.1. Problem

How would you set up an experiment to measure the pair production cross section for the 6.13-MeV N^{16} γ radiation?

33.2. Solution

33.2.1. Introduction

The radioactive isotope N^{16} has two dominant beta energies at 4.26 and 10.4 MeV. There exist also characteristic high energy γ lines at 6.13 and 7.12 MeV. Numerous reaction processes lead to the isotope N^{16}. List at least six nuclear reactions which can be used to produce N^{16} using relatively low energy incident projectiles. The most effective way to produce N^{16}, which has a half-life of 7.35 sec, is by the (n, p) process with O^{16}. In ordinary water-cooled nuclear reactors O^{16} is present, in the form of H_2O, in the strong fission neutron field. We can therefore get a strong N^{16} source by using the water which passes through the core of the operating reactor. The cross section for fission neutrons is rather small (0.019 mb). Nevertheless very strong activities are obtainable because of the enormous intensity of neutrons in the fuel region. In many cases the N^{16} activity is so high that special precautions are necessary to reduce the N^{16} activity. Water running through the reactor core for cooling purposes passes through a labyrinth passage holdup tank causing it to remain at least ten half-lives in a shielded room before it is allowed to proceed to the heat exchangers and return to the core. (By what factor is the N^{16} activity reduced in such a holdup process?) If a water-cooled nuclear reactor is available, we only have to introduce a special collimator somewhere in the core exit region in the cooling system to get an appropriate source of the 6.13–MeV γ radiation.

33.2.2. Pair Production

The Bethe–Heitler theory[1] of pair production is based on the Born approximation but does not agree very well with experimental results in

[1] H. H. Bethe, and W. Heitler, *Proc. Roy. Soc. (London)* **146A,** 83 (1934).

* Problem 33 is by M. J. Higatsberger.

the energy range above the pair production limit, as well as for materials of high Z values. There exists a modification of this theory by Bethe and Maximon[2] which covers the high energy ranges well. For the lower energy range, i.e., approximately $2\,m_e c^2$, a treatment by Jaeger and Hulme[3] has been published which avoids the Born approximation. For the energy range from 3 to 10 MeV numerical results[4] based on the Bethe–Heitler theory are available. This energy range is of particular interest for shielding calculations for nuclear reactor systems. In the range of 6 MeV only the pair production and Compton effects make important contributions. Estimate the relative magnitude of the cross sections for these two processes in this energy range. Show that the photoelectric cross section is negligible in this energy range.

For the determination of pair production cross sections the transmission method is in general use; see, for instance, the work published by Colgate.[5] In the particular case of 6.13-MeV pair production cross sections, a modification is required in the measurement of the relative transmission values for different substances. Show that if the samples are chosen in such a way that the same number of electrons is always presented to the incident photon beam, the Compton cross section cancels in this transmission experiment. Therefore, by measuring the relative transmission values, find an equation for the ratio of the pair production cross sections of the two substances under comparison.

33.2.3. Measurements

Figure 33.1(PE) is a diagram of a sodium iodide pair spectrometer which may be used for the transmission measurement. With a coincidence system it is possible to register the pair production events in the sample by measuring the 511-keV annihilation quanta. Figure 33.1(PE) shows a possible arrangement. Only if the probability of annihilation for the produced positrons is the same in both samples, are the previous relative measurements correct. According to Staub and Winkler[6] this is possible if the thickness of the samples under investigation is approximately the same. If, for instance, Cu and Ni are compared, the densities are close enough so that there are no differences in the thicknesses. If we are interested in Ag, we must approximate the condition of the same thicknesses by using powdered silver pressed to the right thickness. If a 0.5-mm Cu sample

[2] H. H. Bethe, and L. C. Maximon, *Phys. Rev.* **93**, 768 (1954).

[3] J. C. Jaeger, and H. R. Hulme, *Proc. Roy. Soc.* (*London*) **153A,** 443 (1936).

[4] C. M. Davidson, and R. D. Evans, *Rev. Mod. Phys.* **24,** 79 (1952).

[5] A. Colgate, *Phys. Rev.* **87,** 592 (1952).

[6] H. Staub, and H. Winkler, *Helv. Phys. Acta* **27,** 223 (1954).

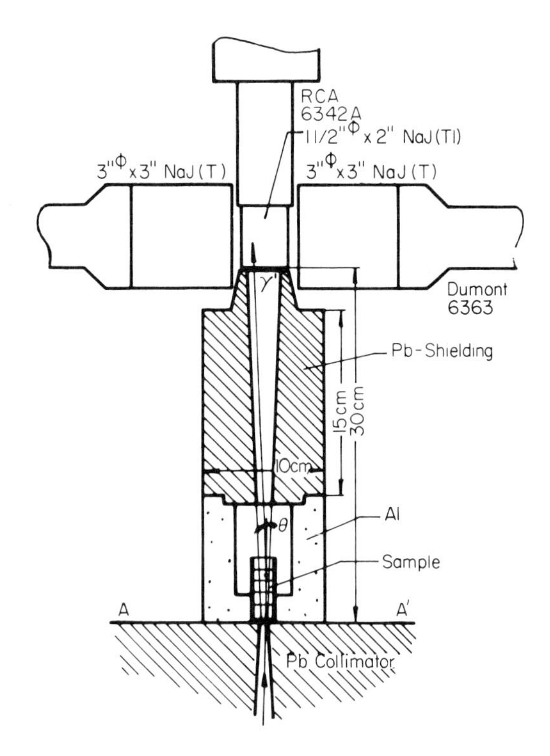

FIG. 33.1(a). NaI pair spectrometer.

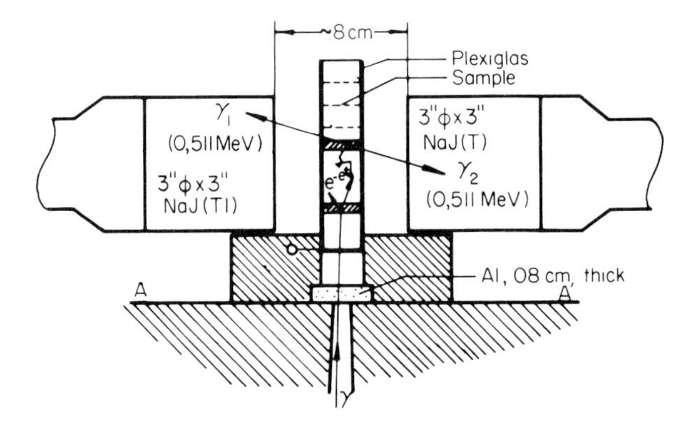

FIG. 33.1(b). Sample arrangement for pair spectrometer.

were used, what thickness in milligrams per square centimeter Ag sample should be used? Figure 33.2 shows actual experimental results compared with theoretical values.[7]

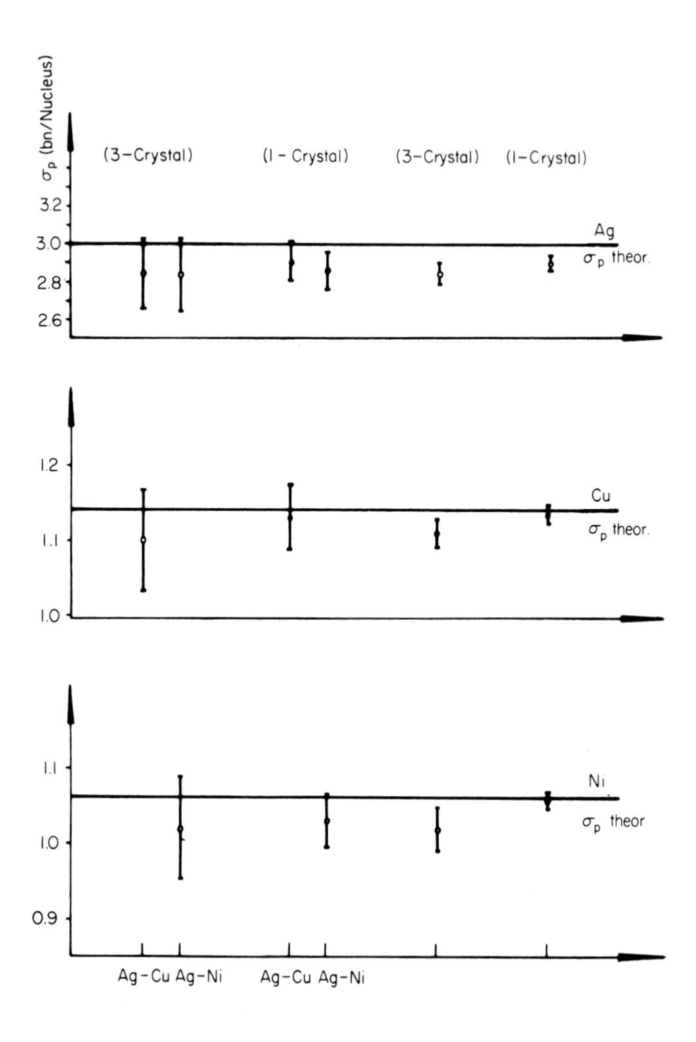

FIG. 33.2. Combined (left-hand side) and transmission (right-hand side) methods; experimental versus theoretical results.

[7] Hae-Ill Bak, and P. Weinzierl, *Acta. Phys. Austriaca* **22,** 1–4, 60 (1966).

33.3. Exercises

33.3.1.

What commercially available electronic units would you use to get at least a 5×10^{-8}-sec time resolution?

33.3.2.

What provisions would one have to make to ensure a constant N^{16} γ activity?

34. REACTIVITY OSCILLATOR*

34.1. Problem

For a research reactor of the swimming pool type construct a simple reactivity oscillator and describe an experimental program to measure the transfer function.

34.2. Solution

34.2.1. Introduction

The dynamic behavior of a nuclear reactor may be described with the aid of the transfer function.[1] The transfer function is a frequency-dependent relation between input and output signals. Such a function is particularly suited for stability investigations. It may also be used in the analysis of other feedback relations in power reactor systems. Normally highly advanced analytical and graphical methods must be applied for interpreting the results. In nuclear reactor technology a basic problem[2] is the determination of the transfer function (power/reactivity) for zero power as well as for design power level. If temperature effects are neglected the feedback mechanism of reactor power to reactivity may be found directly.

34.2.2. Time Dependence of a Zero Power Reactor

The time dependence of a zero power reactor may be described by the following basic equations:

$$\frac{dn}{dt} = \frac{\rho - \beta}{l} n + \sum_{i=1}^{5} \lambda_i c_i \tag{34.1}$$

and

$$\frac{dc_i}{dt} = \frac{\beta_i}{l} n - \lambda_i c_i, \tag{34.2}$$

where $n = n(t)$ is the reactor power; $\rho = \rho(t)$ is the reactivity, $\Delta k/k$; $l = l^*/k_{\text{eff}}$ is the neutron generating time; l^* is the effective neutron

[1] Schulz, "Nuclear Reactors and Power Plants." McGraw–Hill, New York, 1966.
[2] J. M. Harrer *et al.*, *Nucleonics 10* **32** (1952).

* Problem 34 is by M. J. Higatsberger.

lifetime; β_i is the portion of delayed neutrons of group i; β is the sum of delayed neutrons; c_i is proportional to the original concentration of substance for group i neutrons; and λ_i is the decay constant.

Unfortunately the nonlinearity of the system is due to the product ρ. If we restrict ourselves to small sinusoidal changes, we may write $\rho(t) = \rho_1$ sin ωt

$$n(t) = n_0 + n_1 \sin(\omega t + \eta). \tag{34.3}$$

If $\rho_1 \ll 1$ and $n_1 \ll n_0$, show that the transfer function can be written in the form

$$G(j\omega) = \frac{n_1/n_0}{\rho_1} e^{jn} = \frac{1}{j\omega[1 + \sum_{i=1}^{6} (\beta_i/\lambda_i + j\omega)]}. \tag{34.4}$$

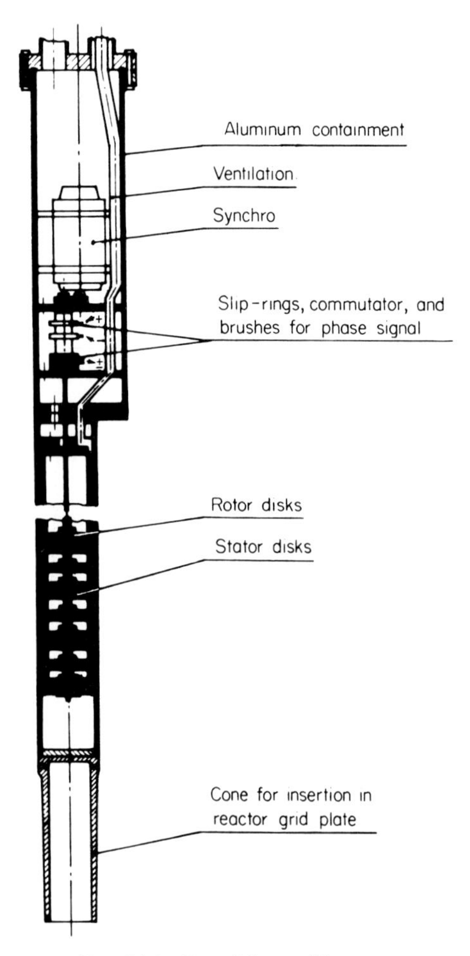

FIG. 34.1. Reactivity oscillator.

34.2.3. Reactivity Oscillator

A means must be provided to change the reactivity periodically, which will result in oscillations in the reactor power. The simplest approach is to introduce into the reactor core an oscillating neutron absorbing media. It is important to recognize that the nuclear reactor is a nonlinear system. Therefore the linear-dependent transfer function may only be applied in the small region around the point under investigation.

34.2.4. Construction Details of the Oscillator

For a particular reactor core under investigation the neutron lifetime l^* was measured to be 65 μsec. It is safe therefore to use an oscillator frequency of $\omega = 500 \sec^{-1}$. Why? But as it is known that temperature effects occur at low frequencies, the other limit of the input signal is set at $\omega = 0.05 \sec^{-1}$. A synchronous motor or mechanical means are therefore required to vary the oscillator frequency over a range of 4 decades. In Fig. 34.1 a possible arrangement as built by Nedelik[3] is shown. The oscillator is introduced into a reactor core position (see block diagram of Fig. 34.2).

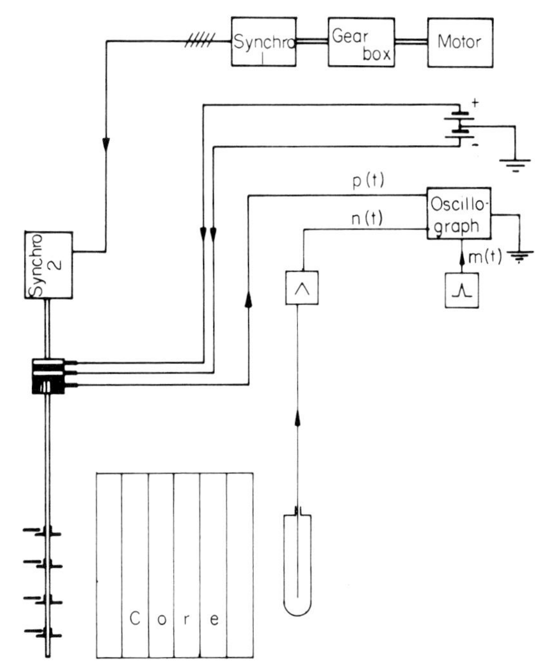

FIG. 34.2. Block diagram where $p(t)$ represents the phase indication, $n(t)$ the neutron signal, and $m(t)$ the timing pulses.

[3] Nedelik "Messungen der ASTRA–Übertragungsfunktion." Isotope in Industrie und Landwirtschaft, Heft 3, 1965.

The oscillator itself is composed of a stator and a rotor, schematically shown in Fig. 34.3. Show that the design indicated in Fig. 34.3 ensures a sinusoidal variation of neutron absorption (i.e., show that the area of the absorbing media rotor plus stator vary sinusoidally) when the rotor is driven at a constant frequency. Stator and rotor are constructed out of Al plates covered with Cd sheets. For cooling purposes at power operation a special venting system is desirable. As the measurement of the phase relationship is very important, an exact indication of the position of the rotor relative to the stator is necessary.

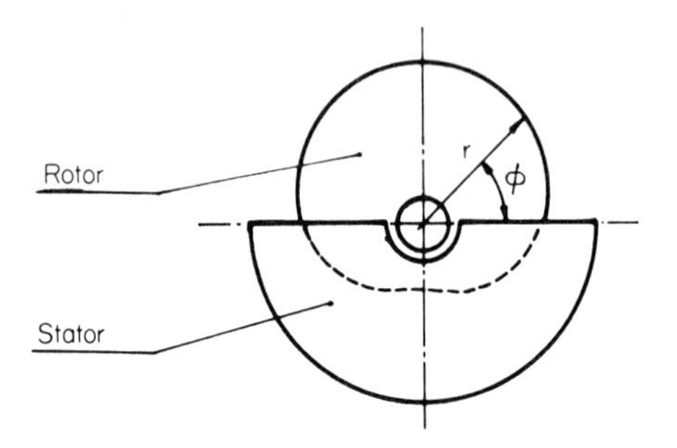

FIG. 34.3. Shape of Cd absorbers; $r = r_0(k+\cos\phi)^{1/2}$.

34.2.5. Results

A given reactivity signal is measured with the help of the fine-control-rod position indicator. For the measurement of power signals a special neutron detecting chamber should be used. A boron-coated ionization chamber near the reactor core may be used, for instance. Describe how this chamber operates to detect neutrons. Can you suggest other types of detectors that could be used for this purpose? It is important to compensate for or shield out the heavy γ radiation. In Figs. 34.4 and 34.5 typical transfer function measurements for amplitude and phase at zero power are compared with a theoretical curve. The difference of a power curve $P(j\omega)$ to the zero power frequency $G(j\omega)$ permits the calculation of the feedback function $H(j\omega)$. What is the relation connecting $H(j\omega)$ with $P(j\omega)$ and $G(j\omega)$?

34.3. Exercises

34.3.1.

Make a design change to the reactivity oscillator shown in Fig. 34.1 so that the rotor position is indicated by a radioactive isotope instead of an electrical signal received from a collector attached to the rotor.

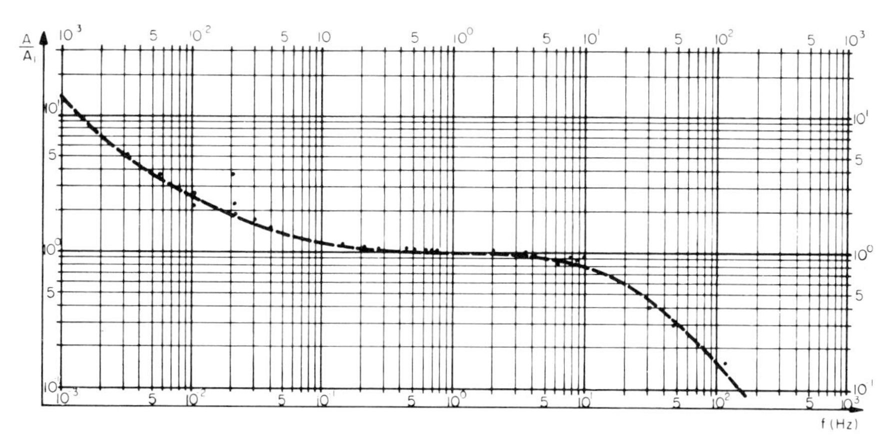

FIG. 34.4. Zero power transfer function—amplitude; the dots represent the measured values, the dashed line the calculated values.

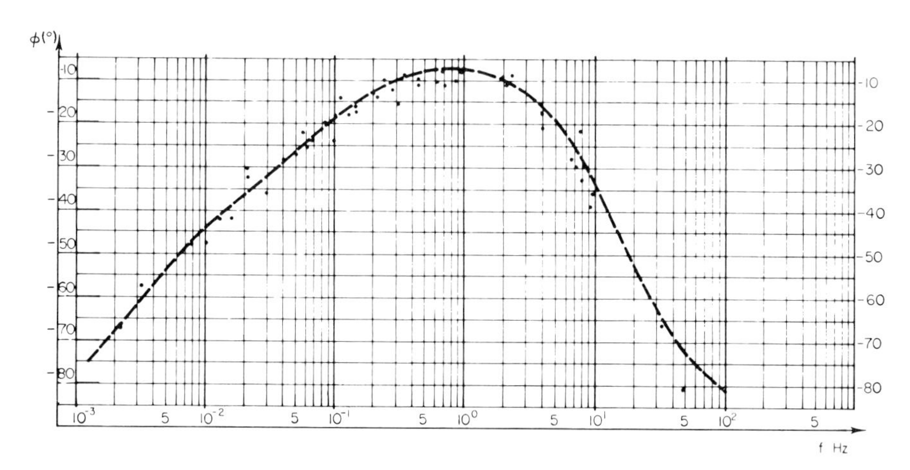

FIG. 34.5. Zero power transfer function—phase; the dots represent the measured values, the dashed line the calculated values.

34.3.2.

How is the stability of the reactor influenced by introducing the temperature and void dependence of reactivity ρ?

35. NEUTRON TRANSMISSION MEASUREMENTS*

35.1. Problem

The concentration of Pu^{239} in mixed nuclear fuel is to be measured by the neutron transmission technique. Elaborate on the feasibility and technical principles of such a method.

35.2. Solution

35.2.1. Introduction

It is well known that the absorption and scattering cross sections for neutrons are dependent on the neutron energy. For low energies characteristic resonance absorption cross sections have been observed which vary from material to material. Relatively accurate values of absorption cross sections have been measured for all fissile materials. These cross section characteristics can be used for quantitative determinations of the relative concentrations of fissile materials, particularly the concentration of U^{235} and Pu^{239}.

35.2.2. Basic Principle

In Fig. 35.1 the total microscopic cross section of Pu^{239}, Sm, and In are given. We note that the characteristic resonance absorption cross section for Pu^{239} has a peak of about 4000 b at a neutron energy of $\sim$ 0.3 eV. At 3 eV, however, the absorption cross section is reduced by a factor of 200. U^{235} does not have a resonance at this same energy. Therefore, if neutrons of the resonance energy are passing a sample of fissionable Pu^{239} a characteristic number of neutrons will be absorbed. If plutonium is replaced in the sample by the same quantity of U^{235}, a completely different number of neutrons will be absorbed. The experimental setup should therefore include a thermal-neutron-energy selector, either a crystal using the Bragg reflection and/or a time-of-flight technique with velocity selectors to obtain a well-defined energy interval. Putz *et al.*[1] have shown what a

[1] F. Putz, T. Al-Khafaji, and P. Weinzierl, Determination of Pu-239 content in nuclear fuel by neutron transmission technique, Nuclear Materials Management, IAEA, Vienna, 1966.

* Problem 35 is by M. J. Higatsberger.

typical setup looks like. Suppose a slab containing Pu^{239} is placed in a well-collimated beam of neutrons. The intensity of the neutrons is measured by a suitable detector after passing through the sample. For the remainder of this derivation the detector is assumed to be a Pu^{239} fission chamber.

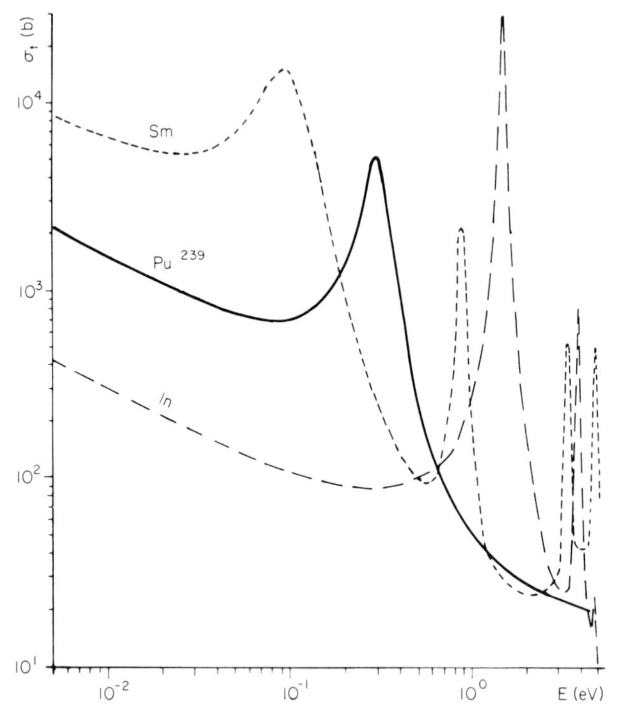

FIG. 35.1. Total microscopic cross section of Pu^{239}, Sm, and In.

Its response to incident neutrons, the energy of which lies between E_1 and E_2, is expressed by

$$R \simeq A \int_{E_1}^{E_2} \Phi(x_D, E)\sigma_f^{49}(E)\, dE \tag{35.1}$$

where $\Phi(x_D, E)$ is the energy-dependent flux at the position x_D of the detector and $\sigma_f^{49}(E)$ is the microscopic fission cross section of Pu^{239}. The proportionality factor A depends on the parameters of the detector. For the flux $\Phi(x_D, E)$ we find

$$\Phi(x_D, E) = \Phi_0(E) \exp[-\int_0^t \{N^{49}(x)\sigma_t^{49}(E) + \sum_i N^i(x)\sigma_t^i(E)\}\, dx], \tag{35.2}$$

where $\Phi_0(E)$ is the unperturbed flux of the incident neutrons on the surface of the sample at $x = 0$; $N^{49}(x)$ is the number of Pu^{239} atoms per unit volume; $N^i(x)$ is the number of atoms of the ith kind contained in the sample per unit volume; $\sigma_t^{49}(E)$ is the total microscopic cross section of Pu^{239}; and $\sigma_t{}^i$ is the total microscopic cross section of the ith kind of atoms.

In deriving formula (35.2) the assumption is made that a collision within the sample leads to the removal of neutrons from the beam. We introduce the abbreviations

$$C^{49}(t) = \int_0^t N^{49}(x)\,dx, \qquad C^i = \int_0^t N^i(x)\,dx \tag{35.3}$$

where the quantities C^{49} and C^i are the total numbers of the corresponding kind of nuclei which are within a cylinder having a unit cross section and a length t parallel to the x direction. Then, show that Eq. (35.1) becomes

$$R = AJ = A\int_{E_1}^{E_2} \Phi_0(E)\exp\{-[C^{49}\sigma_t^{49}(E) + \sum_i C^i\sigma_t{}^i(E)]\}\sigma_f^{49}(E)\,dE. \tag{35.4}$$

Simplify this expression to give the response ($R_0 = AJ_0$) of the detector to the unperturbed neutron beam. Then the transmission $T = R/R_0$ can be written

$$T = \frac{J}{J_0}. \tag{35.5}$$

If the spectrum of the neutron beam and the concentrations of all the materials contained in the sample except that of Pu^{239} are known, the transmission T may be calculated as a function of C^{49} by numerical computation of J and J_0. Then for a measured value of T the corresponding value of C^{49} may easily be determined. In practice the situation is more complex since in an irradiated fuel, for instance, the concentration of the fission products are unknown in general. In this case an approximation must be made. One possible way to simplify the problem is to assume that all the materials together, except Pu^{239}, give rise to the same effect as a single fictitious $1/v$ absorber (where v is neutron velocity). This involves the replacement of the term $\sum_i C^i\sigma_t{}^i(E)$ by $C^f/\sqrt{E}$. Simplify Eq. (35.4) under this assumption. Calculations show that the sensitivity of the transmission T to relative changes $\Delta C^{49}/C^{49}$ is not very high for small values of C^{49} but increases with increasing concentration of Pu^{239}. The transmission increases more rapidly the smaller the energy interval $(E_2 - E_1)$, including the resonance peak at 0.3 eV.

Since the transmission T now depends on two variables, C^{49} and C^f, an additional measurement is needed to determine the Pu^{239} content of the sample. It appears promising to perform this second transmission measurement with well-thermalized neutrons having a Maxwellian energy distribution. In this region the absorption cross section of Pu^{239} also shows nearly a $1/v$ behavior and therefore the dependence of the transmission on C^{49} and C^{f} will be quite different from the dependence in the resonance region. Is the replacement of all the materials other than Pu^{239} by a $1/v$ absorber a justifiable simplification for thermal neutrons?

35.2.3. Numerical Results

The two integrals J_0 and J were evaluated numerically using an IBM–7040 computer, and the results of these calculations are shown in Fig. 35.2 and 35.3. The curves of Fig. 35.2 represent the transmission T_{th}

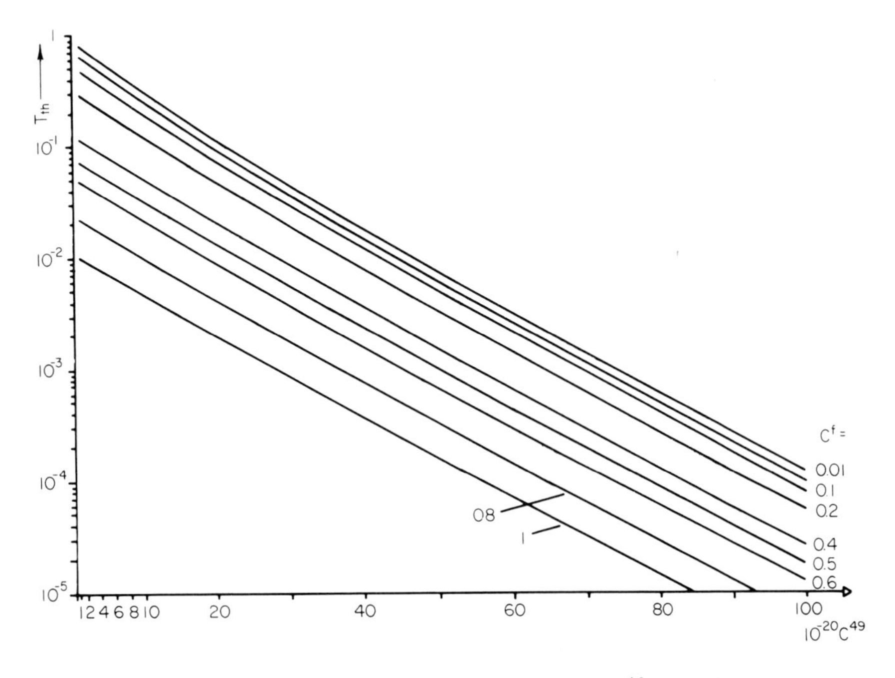

FIG. 35.2. Transmission T_{th} as a function of C^{49} and C^{f}.

for thermal neutrons as a function of C^{49} and C_f. The neutron spectrum was assumed to be Maxwellian and the limits of the integrals were set at $E_1 = 0.001$ eV and $E_2 = 0.5$ eV. Fig. 35.3 shows the transmission T_e in the epithermal energy range as a function of the same variables. In these calculations a $1/E$ flux was assumed and the limits E_1 and E_2 were equal to 0.1 and 0.5 eV, respectively. Neutrons with energies below and above the Pu^{239} resonance at about 0.3 eV should be cut off. To this end the use of a samarium and an indium filter were assumed. Explain why these materials are well suited for this purpose (see Fig. 35.1). If two measured values T_{th} and T_e are given, the corresponding solutions of C^{49} and C^f may easily be determined from the diagrams of Fig. 35.2 and 35.3 by graphical methods.

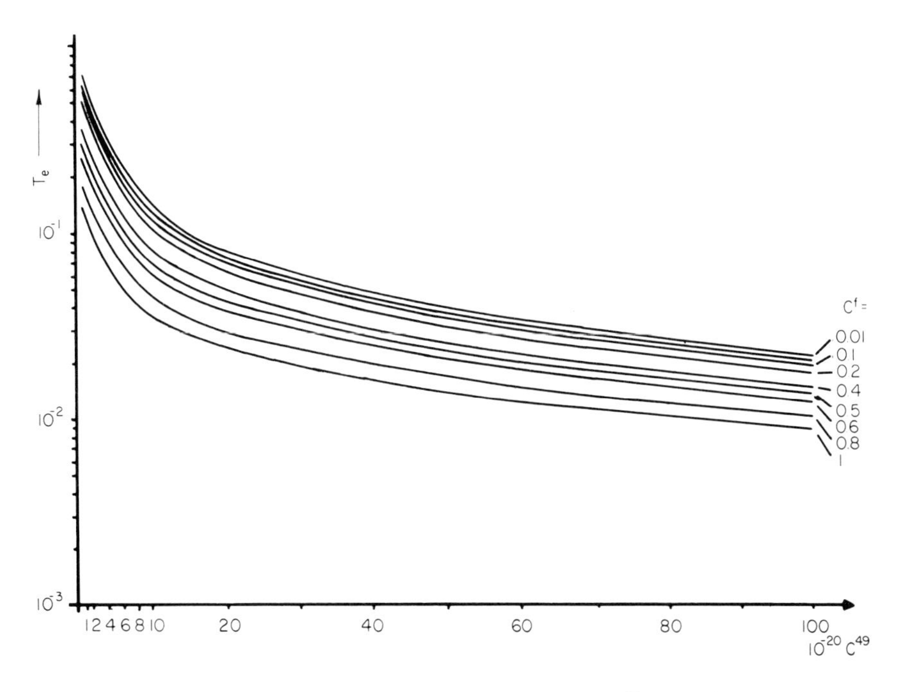

FIG. 35.3. Transmission T_e as a function of C^{49} and C^f.

35.2.4. Final Remarks

The approximations made in deriving the expressions for T may be justified by comparison with experiments. In addition the influence of Pu^{241} must be studied. Pu^{241} unfortunately has a small resonance in its absorption cross section at almost the same energy as Pu^{239} does, but normally the Pu^{239} concentration is dominant over Pu^{241}. Show that for an arbitrary shaped absorber and inhomogeneous distribution of the materials in the absorber it is impossible to give an expression for T which is a function of $C^{49} = \int N^{49}(x, y, z)\, dx\, dy\, dz$. The neutron beam, therefore, must be small enough so that the spatial dependence on the coordinates perpendicular to the direction may be neglected.

35.3. Exercises

35.3.1.

In conversion fuel the amount of Pu^{239} and U^{235} may be comparable, but fission product resonance absorption will play a marked role. Elaborate on this case.

35.3.2.

Design a combination of neutron energy selection for 0.25 ± 0.05 eV based on a combination of mechanical selection followed by crystal deflection. Estimate the intensity relation with a neutron beam coming from a typical water-cooled and moderated reactor.

36. FUEL IRRADIATION CAPSULE*

36.1. Problem

In connection with irradiation tests required for a high temperature reactor fuel (uranium thorium carbide in the form of small pyrolitic coated particles) a static capsule is needed which is capable of irradiating the fuel at a temperature well above 1000°C. To reach the desired temperature only the fission heat of the fuel under test may be used. Indicate design parameters for different materials and discuss the most important safety aspects.

36.2. Solution

36.2.1. Introduction

Before any newly developed type of nuclear fuel is used in large reactors it is subject to several stringent tests. The most important information required is its burnup behavior. Research and test reactors are therefore often used as strong neutron sources for such burnup and irradiation studies. Loop experiments as well as static capsule irradiations are in use. In the lower temperature region Al capsules are commonly in use. For high temperature irradiation tests, graphite has so far been the dominant construction material. For the range above 1000°C only a very few materials are suited from the point of view of heat resistance and neutron absorption cross section. Actually, besides graphite only niobium is worth consideration. Niobium has an absorption cross section of about 1.1 b; that of carbon is 0.0037 b. Any of the other high temperature materials like tantalum and tungsten have cross sections more than one order of magnitude higher than that of niobium so that any advantage in structural behavior may very well be outweighed by flux depression.

36.2.2. Design Consideration

Irradiation test capsules which are to be used in water-cooled-type reactors must be especially designed from the safety point of view. Temperatures above 1000°C will involve substantial hazards. Capsules with a

* Problem 36 is by M. J. Higatsberger.

power density leading to a fuel temperature of 1400°C, for instance, must have a helium barrier thickness of only 0.6–0.8 mm. Then the heat produced in the fuel under test can be carried off properly to the cooling water which is at room temperature. The temperature gradient in such a system is almost 2000°/mm. Slight changes in gap distance may increase the ratio of the fuel temperature to the cooling wall temperature to unsafe conditions. When niobium is used on the construction material, the helium gas can be partly replaced by Al_2O_3. Advantage can be taken of this fact and in Fig. 36.1 the basic design of the graphite capsule is compared with

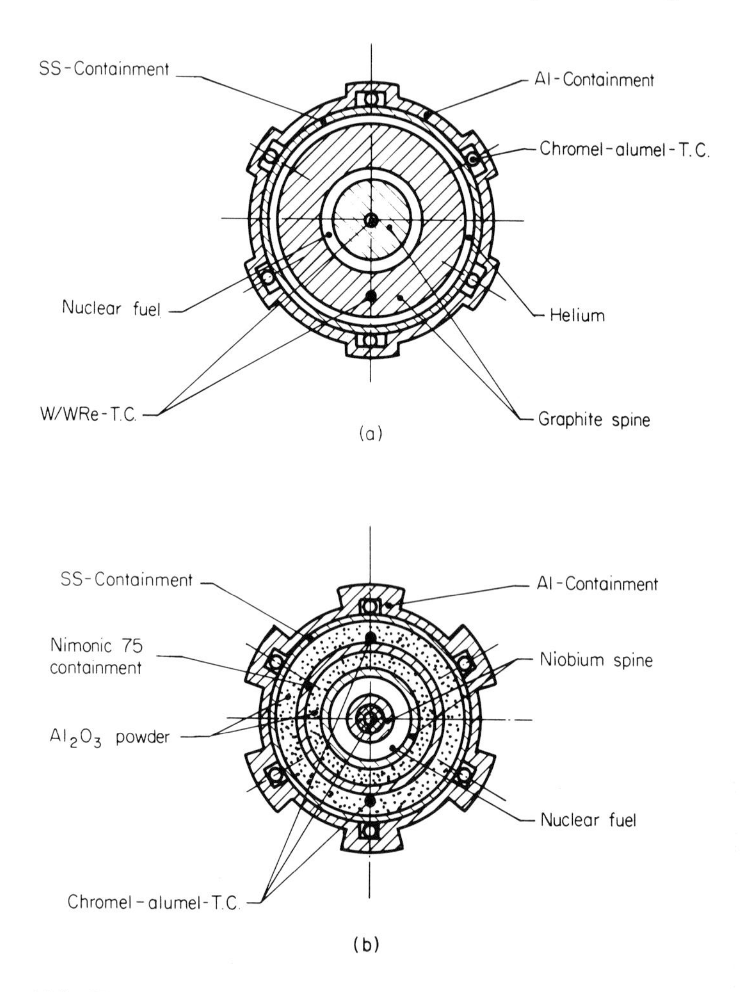

FIG. 36.1. High temperature irradiation capsules: (a) graphite; (b) niobium nimonic Al_2O_3.

that of the niobium capsule. Very little data have been published concerning the behavior of Al_2O_3, but it is evident that the following parameters will be important: conductivity and purity of helium gas, density of the Al_2O_3 powder, and particle size of the powder. Furthermore, at such high temperatures, heat conduction by radiation will not be negligible. A simplification of the capsule construction is possible by taking advantage of the fission heat produced in each fission process. If U^{235} is the fissionable material, show that the heat production in kW is

$$P = \frac{N_{U^{235}} \sigma_f \Phi_d}{3.12 \times 10^{13} \times 2 \times 10^2} G_{U^{235}} = (0.46 \pm 0.017) \frac{\phi_d}{10^{13}} G_{U^{235}} \tag{36.1}$$

where Φ_d is the disturbed thermal neutron flux; $N_{U^{235}} \sigma_f \; 1/V$ is the corrected fission cross section of U^{235} at 100°C; and G is the weight of U^{235} in grams.

An important design parameter is the value of the neutron flux depression. The flux depression due to the capsule material should best be measured. The fission cross section for U^{235} can be simulated with particles of boron and lithium carbide. For typical graphite capsules the flux depression is 0.62 and for the niobium capsule it will be found to be 0.5. At a given constant flux level how much longer must the niobium capsule remain in a given core position in order to lead to the same burnup?

36.2.3. Calculating the Radial Temperature Distribution

The capsules of Fig. 36.1 are designed so that the heat flux is carried off radially. By proper radiation shields even for short capsules the radial heat flux is dominant. We start the calculation by seeking the heat transfer from the outside of the aluminum containment to the cooling water. If q is the thermal power of the capsule per unit length in watts per centimeter and $h = 1.05$ W/cm²/°C is the heat transfer coefficient, show that the thermal gradient for heat transfer from outside the aluminum containment to the cooling water is

$$\Delta_T = \frac{q}{2\pi r h}. \tag{36.2}$$

Show that the thermal gradient for heat conduction through the Al metal housing itself is:

$$\Delta_T = q \frac{\ln(r_1/r_2)}{2\pi \lambda_{\text{metal}}} \tag{36.3}$$

where λ is the thermal conductivity in watts per centimeter per degree centigrade. What is the expression for the thermal gradient for heat conduction through the Al_2O_3 powder? The last information required is the temperature gradient inside the nuclear fuel which we assume to be coated particle fuel. With λ_{Sp} as the thermal conductivity of the fuel spheres show that the thermal gradient is:

$$T = \frac{q}{2\pi\lambda_{Sp}} 0.5 - \frac{r_2^2 \ln(r_1/r_2)}{r_1^2 - r_2^2} . \qquad (36.4)$$

36.2.4. Results

The neutron flux in the center of the reactor core was determined to be:

$$\rho_{th} = (1.81 \pm 0.06) \times 10^{13} \text{ neutrons/cm}^2 \text{ sec/MW}. \qquad (36.5)$$

For the niobium capsule with a sample of 2 g of U^{235}, what is the heat power density from fission?

36.3. Exercises

36.3.1.

Compare the advantages and the disadvantages of the graphite-type capsule with the niobium capsule shown in Fig. 36.1.

36.3.2.

Calculate the time required for 10% burnup of 2 g of pure U^{235} if the reactor flux is 1.8×10^{13} neutrons/cm^2 sec/MW and if the reactor is continuously operated at 5 MW. Calculate also the maximum temperature to which this fuel under test is exposed.

36.3.3.

Describe a simple way of measuring the reactivity change by introducing 2 g of U^{235} with a niobium capsule in a critical reactor.

37. REACTOR REGULATION AND CONTROL*

37.1. Problem

Determine by various experimental methods the reactivity worth of the regulating and control rods for a research or test reactor.

37.2. Solution

For control and safety purposes, as well as for varying the power level, control and regulating devices are required in every nuclear reactor. In general, the control devices are solid, neutron-absorbing materials in the form of rods or plates. The control rods are operated either inside the core or near the reflector region. The control rods or control plates in water-cooled reactors are located normally in interchangeable core positions. The safety of the whole reactor system depends on the amount of reactivity they control. The worth of an individual rod, or a bank of several rods can be measured in several different ways, e.g., the method of the asymptotic positive period, the comparison method, or the rod drop method.

37.2.1. Rod Worth Determination by the Method of the Asymptotic Period

Figure 37.1 shows a typical swimming pool reactor core fueled with highly enriched uranium. The core consists of 14 standard elements, five elements housing control elements, and one element housing the regulating rod. The fission chain reaction is started by the plutonium beryllium source S and the low power counting is done by the fission chambers F_1 and F_2. Describe the principles of operation of a fission chamber and explain why this chamber is used for low power counting? Compensated and uncompensated ionization chambers cover the high power range. Describe the principles of operation of compensated and uncompensated ionization chambers? Why are these detectors used for high power counting? From the core loading one can see that some of the control elements, even if the absorbing material by weight and area is the same, have different reactivity values. So, for instance, the control elements C_1 and C_5 as well

* Problem 37 is by M. J. Higatsberger.

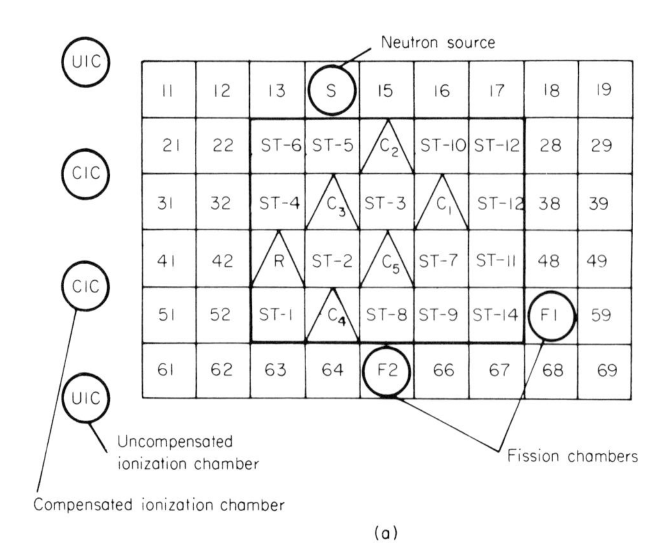

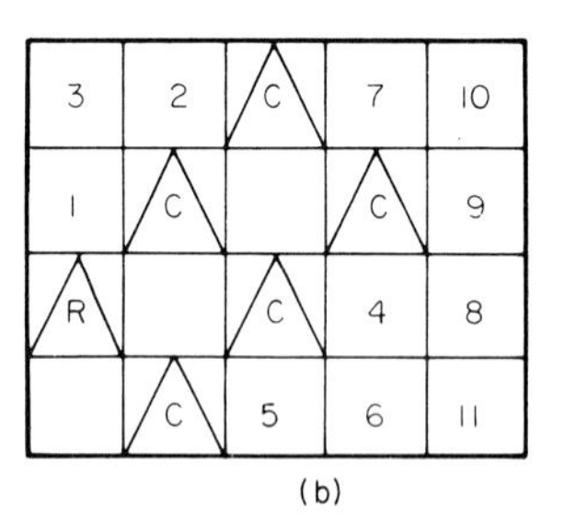

FIG. 37.1. Typical swimming pool reactor core: (a) core loading; (b) loading sequence, where ST is the standard fuel element, C is the control element, and R is the regulating rod element.

as C_3 are more centrally located than the rest of the control elements. The method of the asymptotic period is based on the reactor being in a steady state critical situation when a positive reactivity is introduced. Show that the reactor period T_e is related to the doubling time by

$$T_e = 1.44 \times T_2. \tag{37.1}$$

From the in-hour equation show that the positive period is related to the reactivity by

$$\frac{\Delta K}{K} = \rho = \frac{l^*}{K_{\text{eff}} T_e} + \sum_{i=1}^{i=5} \frac{\beta_i}{1 + \lambda_i T_e} \tag{37.2}$$

where l^* is the effective neutron lifetime; β_i is the portion of delayed neutrons of group i; and λ_i is the decay constant. In Table I the β_i and λ_i for five groups of delayed neutrons are listed.

TABLE I. β_i AND λ_i VALUES FOR DELAYED NEUTRONS

Group	λ_i (sec^{-1})	β_i (10^{-5})
1	0.0125	25
2	0.0315	166
3	0.154	213
4	0.455	241
5	1.61	85

37.2.2. Comparison Method

If one rod's worth with respect to the depth of insertion in the core is known, one can calibrate a second rod by withdrawing it a certain amount, and observing how far the calibrated rod must be inserted to restore criticality. Unfortunately, in small reactor core volumes this method is not very accurate because shadowing effects take place with resulting poor accuracy.

37.2.3. Rod Drop Method

The control rod worth measurement by this method has two advantages. First of all, it measures the integral worth at once and secondly this method may be applied, even if the excess reactivity of the core is small. The procedure is as follows: the reactor is brought to criticality with the rod to be measured completely withdrawn. By a special arrangement the magnet current for the rod under investigation is interrupted, causing the control rod to drop into the core, leading to a shut down of the reactor. From the kinetic equation of reactor dynamics show that the time dependence of the flux for such a sudden negative reactivity change is given by:

$$\Phi_{(t)} = \Phi_0 \frac{1}{\beta + \Delta_K} \sum_{i=1}^{i=5} \beta_i \exp(-\alpha_i t)[1 - \exp(-\lambda_i T)]$$

$$\text{with} \quad \alpha_i = \lambda_i \left(\frac{1 - \beta_i}{\beta + \Delta_K} \right), \qquad (37.3)$$

where β is the sum of all β_i; Φ_0 is the flux at the fission chamber for the critical reactor; T is the time, while flux Φ_0 existed; and t is the time after the negative reactivity input. Because proportionality between flux and count rate exists, we measure the counts per second before and after inserting the negative reactivity. Show that the ratio of the counts per second before, to the total counts measured for a time from t_0 to t_1 after inserting the negative reactivity, is given by (for T sufficiently large)

$$R = \left\{ \frac{1}{\beta + \Delta_K} \sum_{i=1}^{i=5} \frac{\beta_i}{\alpha_i} [\exp(-t_0\alpha_i) - \exp(-t_1\alpha_i)] \right\}^{-1}. \qquad (37.4)$$

37.2.4. Measurements

The asymptotic period measurement is the simplest. The reactor is brought to criticality and then the rod under investigation is withdrawn a certain amount. The positive period or doubling time is measured and $\Delta\rho$ can be immediately calculated for the distance through which the control rod was withdrawn. The comparison method is self-explanatory, if one integral rod curve is known. The rod drop method, finally, consists of a comparison of the count rates at criticality and after the negative reactivity is inserted.

37.3. Exercises

37.3.1.

Derive from the neutron diffusion equation the in-hour formula.

37.3.2.

Discuss the shadow effect of control rods and give examples where good use is made of this effect.

38. ISOCHRONOUS REACTION SURFACES*

38.1. Problem

38.1.1. General

Consider the general nuclear reaction involving an accelerated particle of mass M_1 and laboratory energy E_1 striking a stationary target nucleus M_2 resulting in fragments M_3 and M_4 with a certain reaction Q. Using the kinematic conservation laws show that the isochronous reaction surface for fragment M_3 (i.e., the surface on which the particle M_3 arrives with a fixed elapsed time independent of the emission angle) is a sphere of radius R centered a distance S from the target on the beam line extended with the ratio R/S given by

$$\frac{R^2}{S^2} = \frac{M_4}{M_1 M_3}\left[M_2 + (M_1 + M_2)\frac{Q}{E_1}\right]. \tag{38.1}$$

Also, show that the flight time of fragment M_3 from the target to any point on such a sphere is

$$t = \frac{S(M_1 + M_2)}{M_1 v_1}. \tag{38.2}$$

Justify the statement that any arbitrarily shaped area detection system for the particle M_3 if moved along the isochronous spherical surface (i.e., rotated about the sphere's center at a radius R) while subtending different laboratory solid angles, has a constant center-of-mass solid angle.

38.1.2. Application to Fast Neutron Spectroscopy[1]

Consider the possibility of devising a neutron spectrometer based on applying the above considerations to the *n-p* scattering process and using time-of-flight techniques. Such an idealized isochronous time-of-flight spectrometer might be visualized as in Fig. 38.1. It would involve the use of two hydrogenous scintillators, the first one acting as both the scatter and time zero reference with the second scintillator having a large surface area conforming to the isochronous sphere and indicating the detection

[1] M. L. Roush *et al.*, *Nucl. Instr. Methods* **24**, 290 (1963).

* Problem 38 is by W. F. Hornyak.

time of the scattered neutron. The required photomultiplier tubes are not shown.

Show from Section 38.1.1 that for the n-p scattering process $R = S$, as shown in Fig. 38.1. Also show that the contribution to time resolution due to the shell thickness of the stop detector will be independent of the particular scattered neutron path (angle θ) if the shell scintillator conforms physically to the space between two spheres of radii R and $R + \Delta$, tangent at the start scatterer location [i.e., centers at S and $S + (\Delta/2)$].

Discuss the interception efficiency for the scattered neutrons n', with θ_1 and θ_2 defining the ends of the stop detector. In this, consideration

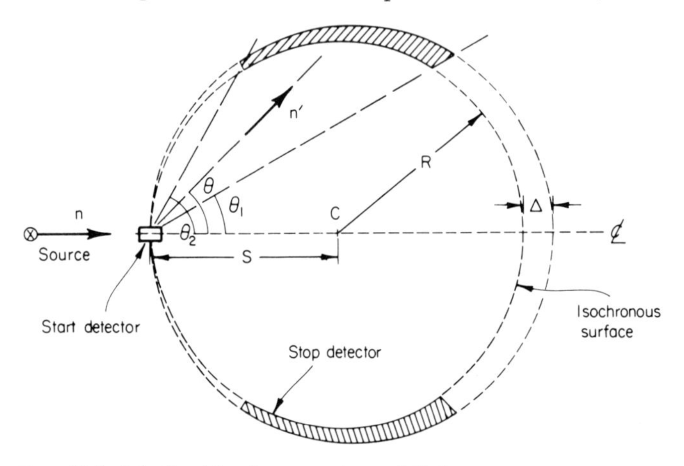

FIG. 38.1. Idealized isochronous time-of-flight neutron spectrometer.

must be given to the variation of neutron detection efficiency with energy for organic (i.e., plastic) scintillators. Why? Determine from a search of the literature the detection efficiency variation with pulse-height-biased arrangements.[2, 3]

Discuss the effects on the resolution of the spectrometer due to the finite length and diame*er of the start detector, taken to be cylindrical in shape. In this connection consideration should be given to variations in transit time between start and stop detection as well as multiple scattering in the start detector.

Discuss the effects on resolution of the phosphor decay time, light collection efficiency (particularly in the stop shell detector), and light transit time spread. In this connection consider methods for collecting the light in the stop detector. Include the possibility of placing a phototube at C, the

[2] J. B. Birks, "The Theory and Practice of Scintillation Counting," Pergamon Press, Oxford Univ. Press, London and New York, 1964.

[3] L. C. L. Yuan and C-S. Wu, eds., "Methods of Experimental Physics," Vol. 5A. Academic Press, New York, 1961.

center of the isochronous sphere, with a reflecting coating on the outside of the shell. Why would this be an arrangement worth considering? Discuss the selection of the scintillator material, referring to current literature.[2-4] Give consideration to the elastic and inelastic scattering of incident neutrons by the elements other than hydrogen in the possible choice for the start scintillation detector. Be sure to consider the appropriate isochronous surfaces for these reactions and estimate the effects with the stop shell designed to match the proper surface for *n-p* scattering. Discuss the effect of Compton scattering of γ rays coming from the source that might also be incident on the start detector.

Discuss the spread in resolution introduced by the photomultipliers and electronics. In this connection discuss the selection of photomultipliers, including the possibilities of using the 56 AVP, 56 AVP/03, 56 AVP/05, or XP1020 for the start detector and the XP1040 or 60 AVP for the shell detector.[2,3,5,6] Search the literature for possible time-to-pulse height converters that might be used and discuss methods of data presentation.

38.1.3. Application to Neutron Polarization Studies.[3,7]

Consider the possibility of designing a combined neutron polarimeter and energy spectrometer to determine the polarization of various neutron energy groups in a collimated beam. A possible arrangement might be as shown in Fig. 38.2. Here the neutrons are first scattered in a helium (He^4) start scintillator and then detected in either a "left" stop scintillator or a "right" stop scintillator (both plastic). The ratio of left to right counting rate determines the polarization while the elapsed time determines the energy. Determine the proper ratio of R/S for elastic scattering from He^4.

If the polarimeter-spectrometer is considered primarily for $2 < E_n < 20$ MeV, what maximum range of θ may be permitted such that the polarization in He^4 scattering is never less than 0.8.[3]

If each of the stop scintillators had a shape determined by the intersection of a right circular cone of half-angle $\frac{1}{2}(\theta_{max} - \theta_{min})$ (the θ's as determined above) and axis $\frac{1}{2}(\theta_{max} + \theta_{min})$ with the isochronous sphere, calculate:

(i) the solid angle of each stop detector (c.m. system)

(ii) the maximum effective depolarization due to the angle taking the scattered neutron out of the plane of Fig. 38.2, while still entering the stop detector.

[4] D. Aliaga-Kelly, and D. R. Nicoll, *Nucl. Instr. Methods* **43,** 110 (1966).

[5] E. Gatti and V. Svelto, *Nucl. Instr. Methods* **43,** 248 (1966).

[6] Photomultiplier Tubes, Philips Electron Tube Division (issued annually, latest October 1965).

[7] Phillips, Marion, and Risser, eds., "Progress in Fast Neutron Physics". Univ. of Chicago Press, Chicago, Illinois, 1963.

If for the detector system designed as discussed above we assume that the effective polarization on scattering from the He^4 is $P_2 = 0.90$, determine an expression for the polarization P_1 in a possible (d, n) process under study when the simultaneous right and left counting rates are N_R and N_L, respectively.

Discuss the design of the He^4 cell and its optical coupling to a photomultiplier tube. Include consideration of a high pressure gas phase as well as a liquid phase. Review the literature to determine what additives could be used to increase the light efficiency.[2] Could scattering of incident neutrons from such additives be distinguished by transit time and pulse height effects from He^4 scattering?

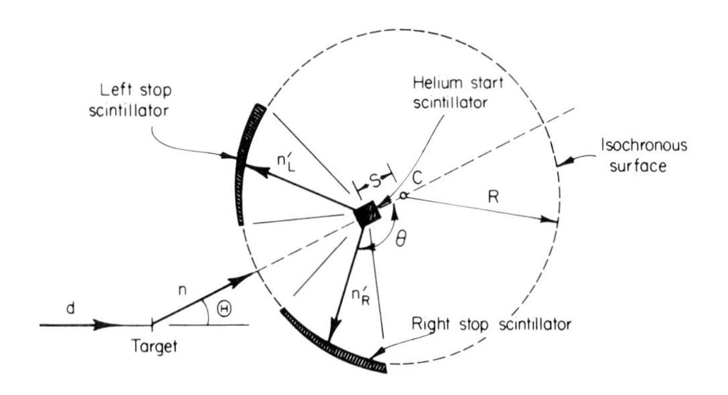

FIG. 38.2. Neutron polarization application of isochronous principle. The source of neutrons is a (d, n) reaction, as an example.

Consider the effects of multiple scattering in the helium cell both on the polarization and time resolution.

How would you determine experimentally if the left and right stop detectors were symmetrically positioned with respect to the primary neutrons and had the same efficiency? Would such possible effects be important?

With all of the above factors in mind, design a practical model of a polarizer-spectrometer, determining the overall efficiency per incident primary neutron ($E_n = 4$ MeV) and estimate the energy resolution pertaining to it.[8]

Review the literature for other possible scintillators to use in place of the helium cell which would be capable of acting as a polarization analyzer. Be sure to include C^{12} (in organic scintillators) as a possibility, particularly

[8] Angular Distributions in Neutron-Induced Reactions, Vol. I, Brookhaven Natl. Lab. 400 (1962).

at the higher energies. In this latter case could neutron scattering from C^{12} be distinguished from the n-p scattering that would also be present?

38.1.4. Application to Heavy Ion Reactions

In heavy ion reactions involving exit channel fragments (M_3 and M_4) of considerably different mass, spectroscopy with particle identification is generally possible. Methods commonly use thick solid state detectors with or without foils to detect the light fragments (the foils are used to differentially shift groups of p, d, α's, etc., that might have the same energy). In more difficult cases "(dE/dx) and E" telescopes may be used.[9,9a]

When the two fragments (M_3 and M_4) are comparable in mass, problems tend to be more severe. In some simple situations (relatively few open exit channels) a thin solid state detector discriminating against the more penetrating lighter particles will suffice in spectroscopic identification. On some occasions the use of two solid state detectors in coincidence (set to the appropriate pair of laboratory angles for M_3 and M_4) with the sum of the pulse heights indicating Q and a small angular acceptance for the two detectors assisting in separating various ambiguous kinematic possibilities results in complete kinematic identification.

Since the case $M_3 \approx M_4$ offers some problems, an alternate means of detection using the isochronous reaction surface concept might be considered. This method would be particularly suitable when $M_3 \approx M_4$ and where a relatively small cross section is involved so that a large solid angle is mandatory.

As an example consider the inelastic scattering of O^{16} ions from a C^{12} target, leaving C^{12} in its first excited state at 4.433 MeV. Suppose that the incident O^{16} ions are produced in a pulsed tandem Van de Graaff accelerator with energy $E_1 = 25$ MeV.

If these O^{16} ions are in the 5+ charge state produced in the gas stripper of the tandem, calculate the terminal voltage.

Calculate the Q value[10] for the $C^{12}(O^{16}, O^{16})C^{12*}$ ($E_x = 4.433$ MeV) reaction and determine $v_{c.m.}$, v_{3c} ($M_3 = 16$), R if $S = 120$ cm, and calculate the transit time of the inelastic O^{16} ions from the target to the isochronous sphere.

Calculate the solid angle in the center-of-mass system for two possible detectors contoured to fit the isochronous spherical surface, having chord diameters of 32 and 54 cm. Calculate the minimum thickness of detector

[9] M. W. Sachs *et al.*, *Nucl. Instr. Methods* **41,** 213 (1966).

[9a] S. K. Mark, and R. B. Moore, *Nucl. Instr. Methods* **44,** 93 (1966).

[10] Consistent set of energies liberated in nuclear reactions, Nuclear Data Project. National Academy of Sciences. (1961).

(plastic NE102A, for example) necessary to stop O^{16} ions of approximately 25 MeV.[11] Discuss the possibility of using other scintillators of comparable thickness. Discuss the selection of a photomultiplier; include consideration of the XP1040 and 60 AVP.[6] Design an optical coupler or light pipe between the thin detection scintillator and the photomultiplier tube face. In order to improve the light collection efficiency at the photomultiplier, discuss the possibility of "aluminizing" the surface of the scintillator facing the target with a coating of 50–100 $\mu g/cm^2$. Would such a reflecting surface also be useful in aligning the isochronous surface with respect to the target and beam axis?

How would you use a specific fabricated isochronous surface detector of radius R if you wished to change either E_1 or Q (i.e., changed S)?

The question of energy resolution for the isochronous system is relatively complex. One aspect could be described as a "Q resolution." For example, if in the reaction $C^{12}(O^{16}, O^{16})C^{12*}$ ($E_x = 4.433$ MeV) the 4.43-MeV level were a doublet, how close could this doublet be and still be resolved in arrival time? In answering this question first derive the general expression pertaining:

$$\Delta Q = \frac{2\,\Delta t}{t}\left(1 + \frac{S}{R}\cos\theta_c\right)\left(\frac{M_2 E_1}{M_1 + M_2} + Q\right) \qquad (38.3)$$

where Δt represents the time resolution of the combined effects of the scintillator, photomultiplier, and subsequent electronics. Calculate ΔQ per nanosecond for $\theta_c = 60°$, the other factors as before. Determine ΔE_3/nsec, the energy resolution a small energy sensing detector subtending a negligible angle would have to have for the same ΔQ/nsec at the same laboratory angle. In the large solid angle application requiring the larger tubes, estimate the time resolution attainable, then estimate ΔQ and ΔE_3. In an application designed to measure angular distributions, assume a chord diameter of 8 cm for the isochronous detector, other dimensions being the same. Calculate the half-angle subtended in the center-of-mass system and the solid angle. If such a detector could be successfully coupled to an XP1020 photomultiplier, estimate the time resolution. Again determine ΔQ and ΔE_3. Calculate $\Delta E_3'$, the kinematic energy spread that an energy sensing detector would have if it also subtended the same center-of-mass solid angle.

The question of resolution pertaining to other open exit channels will be illustrated for O^{16} ions incident on a C^{12} target. If the incident labora-

[11] Studies in Penetration of Charged Particles in Matter, Publ. 1133. National Academy of Sciences (1964).

tory energy of the ions is $E_1 = 25$ MeV, calculate the energy available in the c.m. system. Determine what reactions have favorable Q values. In each case determine the mean time of arrival at the detector using a θ determined by the center of the detector set to give $\theta_c = 60°$ for the inelastic O^{16} ions of interest. In the case of $C^{12}(O^{16}, p)Al^{27}$ estimate the transit time spread for protons intercepted by the extreme edges of the detector. The above kinematic calculations may be done with sufficient accuracy using a graphical addition of the velocity vectors v_3 and $v_{c.m.}$, etc.

The main problem using this method is encountered when ions other than those of interest happen at some Q (leading to some excited state in the exit channel) to have the same value of v_{3c} as the ions of interest. The detector will constitute an isochronous surface for these ions as well. As an illustration calculate the Q value and the kinetic energy in the c.m. system of the exit channel for $C^{12}(O^{16}, p)Al^{27*}$ and $C^{12}(O^{16}, \alpha)Mg^{24*}$ such that v_{3c} is in each case the same as for the desired inelastic O^{16} scattering. As an example of a similar problem arising when one of the fragments is particle unstable leading in all to three final particles in the exit channel, consider the case $C^{12}(O^{16}, Be^{8*})Ne^{20}$ with $E_x = 2.90$ MeV in Be^8. Assume sequential processes with the $Be^{8*} \rightarrow 2\alpha$ decay occurring last. Using a graphical method estimate the angles θ_c at which Be^{8*} could give α particles having the same transit time to the detector as the inelastic O^{16} ions. Is the detector surface again an isochronous surface for these α particles?

Note that in each of these above cases $E_3(\text{lab})/M_3$ is a constant. In view of this could pulse height information from the plastic detector be used to separate these ions?[2, 3] In answering this question consider both the case of a detector thickness greater than the range of the protons and also the case where the detector thickness is equal to the range of 25-MeV O^{16} ions.

As an indication of the probability for these competing reactions, calculate the excitation energy in Al^{27} and Mg^{24} that would be reached. A perhaps better indication could be obtained by comparing the Coulomb barrier penetration ($l = 0$) for the various exit channel combinations at the available energy in the c.m. system in each case. In answering this, show that the barrier height E_B ($l = 0$) can be written as:

$$E_B = \frac{Z_3 Z_4}{1.04(A_3^{1/3} + A_4^{1/3})} \text{ MeV} \tag{38.4}$$

where the effective Coulomb radius has been taken as $R_c = 1.50\,(A_3^{1/3} + A_4^{1/3})$ F and E_B is just the Coulomb energy of two charges $Z_1 e$ and $Z_2 e$ at this separation. Show that the penetration factor may be written as:

$$P = e^{-2\gamma_0}$$

where

$$2\gamma_0 = 0.644\,(A_3{}^{1/3} + A_4{}^{1/3})\left(\frac{Z_3 Z_4 M_3 M_4}{M_3 + M_4}\right)^{1/2}$$
$$\times\,[x^{-1/2}\cos^{-1} x^{1/2} - (1-x)^{1/2}]$$
$$\text{with} \quad x = \frac{E_{\text{c.m.}}}{E_B}. \tag{38.5}$$

Here $E_{\text{c.m.}}$ is the kinetic energy available in the center-of-mass system for the fragments M_3 and M_4.[12] A more correct treatment of the transmission

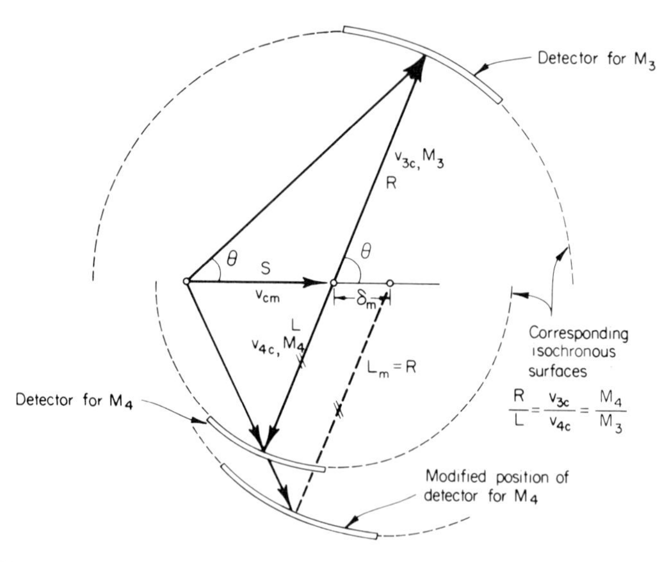

FIG. 38.3. Coincidence arrangement for kinematic pair of particles M_3 and M_4 to achieve mass separation in competing reactions.

coefficients than the relatively naive estimate given here is to be found in the literature.[13, 14]

The use of two isochronous surfaces, one for fragment M_3 and the kinematically corresponding one for fragment M_4 operated in coincidence will largely eliminate the mass identification problems when two reactions have nearly equal v_{3c}. Consider the coincidence arrangement shown in Fig. 38.3 operated with a pulsed accelerator such that $t = S/v_{\text{c.m.}}$ may also be determined.

[12] R. D. Evans, "The Atomic Nucleus." McGraw–Hill, New York, (1955).
[13] T. D. Thomas, *Phys. Rev.* **116,** 703 (1959).
[14] E. W. Vogt *et al.*, *Phys. Rev.* **136,** 99 (1964).

The figure illustrates the case for a desired reaction. Imagine that a competing reaction with $M_3' = M_3 + \Delta M$ and $M_4' = M_4 - \Delta M$ happens to have such a Q value that $v_{3c}' = v_{3c}$. Both reactions give the same value of the transit time to the surface for M_3 and this surface is isochronous for both. However, the surface for M_4 cannot be the isochronous pair for M_4'. Why? Show that while particle M_4 arrives at the surface with radius L in the same time t that both M_3 and M_3' arrive at the surface with radius R, M_4' will strike surface L at a time Δt earlier such that

$$\frac{\Delta t}{t} = \frac{\Delta M(M_3 + M_4)}{M_4[M_3 + \Delta M - (S/R)\cos\theta_c(M_4 - \Delta M)]}. \tag{38.6}$$

This equation could be viewed as a "channel mass resolution" expression since Δt could be considered a function of ΔM when all the other factors are known for the desired reaction.

Consider a hypothetical reaction with $S/R = 1$ and $t = 100$ nsec; show that the most adverse situation (i.e., smallest Δt) results when $\Delta M = 1$ and $\theta_c = 180°$. Calculate Δt when $M_4 = 20$. Also calculate Δt for the same case if $\Delta M = 1$, $\theta_c = 90°$, and $M_3 = M_4 = 20$.

For practical considerations it would be desirable to have two identical detectors for M_3 and M_4. Why? Referring to Fig. 38.3, show that a detector for M_4 identical to that for M_3 may be used if placed in the modified position with L_m (L modified) $= R$, L and L_m parallel, and the center of curvature shifted by δ_m such that $(S + \delta_m)/S = M_4/M_3$. The difference in arrival times of any pair M_3 and M_4 is again independent of θ_c, although not equal to zero as for the pair of surfaces R and L. This simply requires the insertion of an appropriate length delay cable in one leg of the coincident circuit. Why? In this case are the solid angles of the two surfaces correctly matched for intercepting kinematic pairs? If the two surfaces subtended small enough angles to permit angular distribution studies, how should they be moved as θ_c is varied? Need the matching delay cable be altered? If either E_1 or Q is changed will δ_m have to be changed?

Finally, we would like to consider "target mass resolution". When the target contains two masses M_2 and $M_2' = M_2 - \Delta M$, and where for the desired reaction we have M_3, M_4, and R/S (for a particular Q), show that timing is most difficult when

$$\delta = M_4 - M_4' \approx \frac{M_4\,\Delta M}{M_1 + M_2}\left(1 + \frac{S}{R}\cos\theta_c\right). \tag{38.7}$$

In this derivation assume δ and ΔM small compared to all the other masses, and that some Q' exists to make $v_3' = v_3$ for the M_3' called for in the above equation. The various isochronous surfaces are different, hence the time coincidences only result for the particular θ_c.

Show that for elastic scattering (i.e., $Q = Q' = 0$) $v_3' = v_3$ is always (and only) true for $\theta_c = 0$ ($M_3 = M_1$) any ΔM.

38.2. Partial Solutions

38.2.1. General

The diagram shown (Fig. 38.4) pertains to the general nuclear reaction in question. The laboratory velocity and energy of the incident particle are v_1 and E_1, the laboratory and center-of-mass velocity of particle M_3

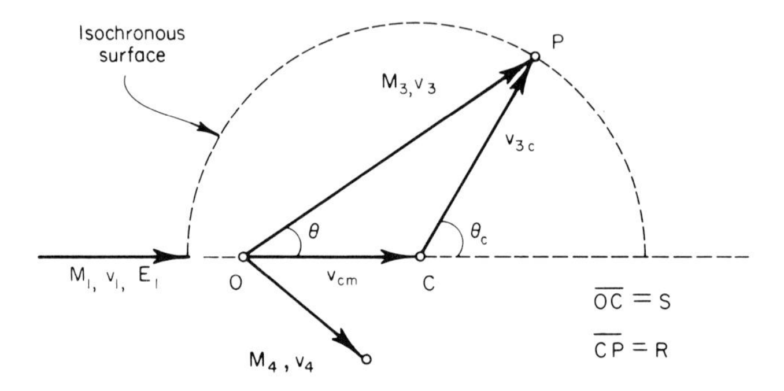

FIG. 38.4. Reaction kinematics illustrating the isochronous surface traced by point P.

are v_3 and v_{3c}, respectively, while $v_{\text{c.m.}}$ is the velocity of the center of mass. Other symbols are self evident.

A trivial application of the kinematic conservation laws gives:

$$v^2_{\text{c.m.}} = \frac{2M_1E_1}{(M_1 + M_2)^2}$$

$$v^2_{3c} = \frac{2M_4[M_2E_1 + (M_1 + M_2)Q]}{M_3(M_1 + M_2)^2}.$$

Clearly $v_{\text{c.m.}}$ and v_{3c} are independent of the emission angle θ_c in the center-of-mass system. It then follows that the fragment M_3 originating its flight at the target position O will arrive at P, a point on a spherical surface with its center at C, in a fixed time independent of θ_c. If the distances $\overline{OC}$ and $\overline{CP}$ are designated as S and R, respectively, it is only required that $R/S = v_{3c}/v_{\text{c.m.}}$ or

$$\frac{R^2}{S^2} = \frac{M_4}{M_1M_3}\left[M_2 + (M_1 + M_2)\frac{Q}{E_1}\right]$$

for the spherical surface of radius R to become an isochronous reaction surface for the particle M_3. The flight time of fragment M_3 from O to P is clearly the same as that for the center of mass to travel from O to C, hence:

$$t = \frac{S}{v_{\text{c.m.}}} = \frac{S(M_1 + M_2)}{M_1 v_1}.$$

38.2.2. Fast Neutron Spectroscopy

To contribute equal effects to the time resolution independent of θ, the stop detector should be confined to fill the region between two isochronous spheres R and $R + \Delta$, with centers separated by $\Delta/2$. This would, of course, give an effective detector thickness of $\Delta \cos \theta$ for the detection of the scattered neutrons. The detection efficiency per incident neutron scattered in the start detector would then vary with θ as $\Delta \cos \theta \; d\sigma(\theta)\eta(n')$ where $d\sigma(\theta)$ is the laboratory differential cross section for n-p scattering and $\eta(n')$ is the biased efficiency per unit length for the plastic stop scintillator at the appropriate scattered neutron energy. For $E_n \lesssim 15$ MeV, $d\sigma(\theta) \approx 2\sigma_T \sin \theta \cos \theta \; d\theta$, and $\eta(n') \approx K/v_n \cos \theta$ for plastic detectors biased to 300–500-keV proton energy equivalent in the range $2 < E_{n'} < 15$ MeV. Thus, the factor $\Delta \cos \theta \eta(n')$ results in essentially constant detection efficiency over the shell per incident scattered neutron. The fraction f of intercepted neutrons in the range $\theta_1 \leqslant \theta \leqslant \theta_2$ is just

$$f = \frac{1}{\sigma_T} \int_{\theta_1}^{\theta_2} \frac{\sigma_T}{\pi} \cos \theta \; 2\pi \sin \theta \; d\theta.$$

Clearly optimal values of f are obtained by taking $\theta_1 = (\pi/4) - \beta$ and $\theta_2 = (\pi/4) + \beta$, resulting in

$$f = \sin 2\beta.$$

Thus, even if relatively large portions of the shell are removed by restricting $\theta_1 = 30°$ and $\theta_2 = 60°$, one-half of all the neutrons scattered at the start detector would be intercepted. What would be the result if $\eta(n') \approx K(v_n \cos \theta)^{-\alpha}$; for example, $\alpha = 2$?

The length of the start detector contributes surprisingly little to the time resolution spread when β is small (i.e. $\theta \approx 45°$) since the path length differences from the front and back portions of the detector to the shell are largely cancelled by the changes in neutron velocity caused by the corresponding changes in scattering angle. Verify that this is so. For collimated neutrons incident on the start scatterer, the spread in resolution introduced by the shell thickness (the factor Δ in Fig. 38.1) and the radius r of the start scatterer is for small β

$$\frac{\Delta t}{t} \approx \frac{1}{R}\left(r + \frac{\Delta}{4}\right).$$

Show that this is so.

Suppose it is decided from such considerations that a practical design might have $R = S = 30$ in., $\Delta = \frac{5}{4}$ in., $\beta = 15°$, and the length and radius of the start scatterer 3 in. and $\frac{3}{16}$ in., respectively. Estimate the fraction of neutrons ($E_n = 2$ MeV) incident on the scatterer that will be scattered at least once. What overall efficiency would the spectrometer have per incident neutron? In making this estimate, assume that all the recoil protons in the start detector produce useable scintillation pulses. Is this reasonable? In the stop shell detector, however, you will have to use an empirical value for the efficiency appropriate for some bias cutoff, selecting only those scintillation pulses giving 5 to 10 photoelectrons at the photoemissive surface of the photomultiplier. Why?

Estimate the energy resolution for $E_n = 2$-MeV incident collimated paraxial neutrons, assuming that a shell detector bias set to 5–10 electrons will give a time spread of 1.5 nsec including the combined effects of phosphor decay time (for example, NE 102 A), light collection efficiency, and light transit time spread.

To determine the effects of multiple scattering consider at least the two following calculations.

(i) If the first scattering of the incident neutron in the start detector was on axis and at $\theta = 45 \pm 15°$, estimate the probability that such a neutron will scatter again from hydrogen before leaving.

(ii) Through how large an angle could a neutron ($E_n = 2$ MeV) scatter from hydrogen and still lose less energy than your estimate of the resolution width above?

38.2.3. Neutron Polarization Studies

Assume the (d, n) process under study results in a polarization P_1 for the neutrons emitted at an angle Θ which are then subsequently scattered by the He^4 cell. The fraction of the neutrons incident on the helium cell with "spin-up" is $(1 + P_1)/2$ and with "spin-down" is $(1 - P_1)/2$. The probability of He^4 scattering the spin-up neutrons to the left (at angle θ) is $\frac{1}{2}(1 + P_2)$ and the probability of scattering the spin-up neutrons to the right (at angle θ) is $\frac{1}{2}(1 - P_2)$, where P_2 is the polarization for elastic scattering of neutrons by He^4 at the energy and angle θ in question. The situation is reversed for the spin-down neutrons. Thus, the ratio

$$\frac{N_L - N_R}{N_L + N_R} = \frac{[(1 + P_1)(1 - P_2) + (1 - P_1)(1 + P_2)] - [(1 + P_1)(1 - P_2) + (1 - P_1)(1 + P_2)]}{[(1 + P_1)(1 + P_2) + (1 - P_1)(1 - P_2)] + [(1 + P_1)(1 - P_2) + (1 - P_1)(1 + P_2)]}$$

or

$$\frac{N_L - N_R}{N_L + N_R} = P_1 P_2$$

hence

$$P_1 = 1.11 \frac{N_L - N_R}{N_L + N_R}.$$

38.2.4. Heavy Ion Reactions

With S taken as 120 cm in the inelastic O^{16} scattering reaction a flight time of 140 nsec results with center-of-mass solid angles of $\approx \frac{1}{25}$ and $\approx \frac{1}{75}$ of a sphere being realizable using the larger photomultiplier tubes.

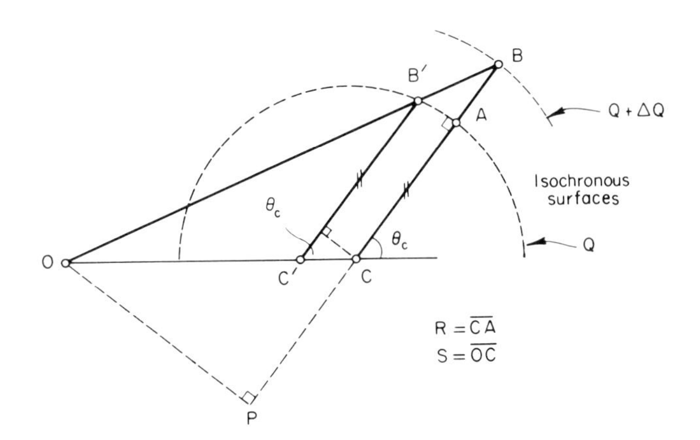

FIG. 38.5. Geometrical relationships for reactions with Q and $Q+\Delta Q$.

The "Q-resolution" equation is readily derived using Fig. 38.5. When the particle $\alpha(Q)$ arrives at point A, particle $\beta(Q + \Delta Q)$ arrives at point B and the center of mass at point C. Particle β was intercepted by the isochronous surface (Q) some time earlier at B' when the center of mass was at C'. If $\Delta Q/Q \ll 1$ then $\overline{B'A} \perp \overline{CA}$, and it follows:

$$\frac{\overline{AB}}{\overline{PB}} = \frac{\overline{B'A}}{\overline{OP}} = \frac{\overline{C'C} \sin \theta_c}{\overline{OC} \sin \theta_c} = \frac{\overline{C'C}}{\overline{OC}} .$$

Thus,

$$\frac{\Delta v_{3c}}{v_{3c} + v_{\text{c.m.}} \cos \theta_c} = \frac{\Delta t}{t},$$

again if $\Delta Q/Q \ll 1$. Since

$$v_{3c}^2 = \frac{2M_4}{M_3(M_1 + M_2)^2} [M_2 E_1 + (M_1 + M_2)Q]$$

$$\frac{\Delta v_{3c}}{v_{3c}} = \frac{\Delta Q}{2[M_2 E_1/(M_1 + M_2) + Q]}$$

therefore

$$\Delta Q = \frac{2\, \Delta t}{t} \left(1 + \frac{S}{R} \cos \theta_c\right) \left(\frac{M_2 E_1}{M_1 + M_2} + Q\right).$$

The main point of the "Q-resolution" and ΔE_3 calculations is to demonstrate the importance of kinematic energy spread for energy sensing detectors. The isochronous detector, of course, does not suffer from this chromatic aberration. In this connection, consider the present state of the art in trying to use a large area solid state detector for timing measurements in place of the scintillator and photomultiplier.

Review the state of the art in pulsing and bunching of tandem accelerators.[15]

[15] L. E. Collins, D. Dandy, and P.T. Stroud, *Nucl. Instr. Methods.* **42,** 206 (1966).

39. NEUTRON THRESHOLD MEASUREMENTS*

39.1. Problem

It is well known that the kinematics of negative Q-value reactions near threshold is relatively complicated. The reaction $Li^7(p, n)Be^7$, $Q = -1644.4$ keV will be taken as an example.[1,2] Precisely at threshold, 29.4 keV energy neutrons appear in the forward direction in the laboratory system ($\theta = 0$). At energies just above threshold, neutrons are emitted only in a small forward cone. The half-angle of this cone increases with incident energy and at 39.1 keV above threshold, 90° and, in fact, the whole backward hemisphere, becomes physically accessible to neutron emission in the laboratory system, with the appearance for the first time of thermal neutrons. Verify these remarks by calculating and plotting E_3 (neutron laboratory energy) as a function of ΔE_1 (incident energy above threshold) with θ (laboratory angle) as a parameter.

The standard procedure for experimentally determining the threshold has been to measure the total neutron yield from a thick target as a function of the incident energy near threshold and to use a linear extrapolation of $Y^{2/3}$ to zero [the assumption here is of dominant s-wave production yielding $Y = K(\Delta E_1)^{3/2}$]. Show that this extrapolation is valid and follows from non resonant s-wave production. What effect would nearby resonances have?

A method which is free of this criticism uses a fixed energy E_1 (set just above the threshold) and determines the maximum angle θ_m in the laboratory system at which neutron emission is kinematically possible. Show that

$$\sin^2 \theta_m = \frac{M_2 M_4}{M_1 M_3}\left(1 - \frac{E_{\mathrm{TH}}}{E_1}\right) = \frac{M_2 M_4}{M_1 M_3}\frac{\Delta E_1}{E_1}. \qquad (39.1)$$

By considering $[\partial(\Delta E_1)/\partial\theta_m]\,|_{E1}$ show that ΔE_1 is determined with greatest sensitivity when $\theta_m = 45°$. Use this to determine the best value for ΔE_1. Then show that the threshold and Q-value sensitivity are about 700 and 600 eV/degree, respectively.

[1] F. C. Young and J. B. Marion, *Nucl. Phys.* **41**, 561 (1963).
[2] R. D. Evans, "The Atomic Nucleus." McGraw–Hill, New York (1955).

* Problem 39 is by W. F. Hornyak.

Examine the possibility of using various slow neutron detectors in the neutron energy range of interest (i.e., $E_3 \approx 10$–30 keV). Be sure to give consideration to physical size, efficiency, and γ-ray sensitivity. If a detector size 3×50 mm is selected with a target to detector distance of 50 cm, estimate the effective angular spread $\Delta\theta$, given that the accelerator and beam focusing devices introduce a target beam spot size of 3 mm and an angular divergence of 0.2°.

Show that the laboratory yield as a function of θ and θ_m at the fixed proton energy in the target E_1 is:

$$Y(\theta, \theta_m) = \frac{K(M_1 + M_2)x^2 \cos\theta}{(2M_1E_1)^{1/2}(1 - x^2 \sin^2\theta)^{1/2}} \tag{39.2}$$

where

$$x^2 = \frac{1}{\sin^2\theta_m}. \tag{39.3}$$

Again, take the reaction to proceed with s-wave production (i.e., isotropic center-of-mass distribution). You will also have to note that at angles $\theta < \theta_m$, neutrons appear in the laboratory system having two different velocities and thus will have to be appropriately weighted by the detector sensitivity. Assume the detector used has a $1/v$ response.

The inclusion of the accelerator beam energy spread and target energy loss is relatively complicated; however, discuss in principle how such a calculation would go [be sure to consider the singularity in $Y(\theta, \theta_m)$ when $\theta = \theta_m$].

The beam spread (full width at half-maximum) of a Van de Graaff accelerator operating at an energy required in this experiment might be 250 eV. Discuss how such a spread could be experimentally measured.

In view of the high precision available in the cone cutoff method, estimate the need for relativistic corrections in the kinematic equations, the need to consider atomic binding and excitation on nuclear Q values,[3] and corrections for Doppler broadening in the target due to thermal motion.[4]

Another example of neutron threshold techniques is in the use of (d, n) reactions leading to excited states in the residual nucleus giving negative Q values. There is reason to believe that a strong yield at threshold is an indication of a large proton reduced width in the residual nucleus for the level in question. A standard way of determining the onset of thermal neutrons indicative of a threshold is to determine the ratio of the yield of thermal neutrons to fast neutrons as a function of incident energy, E_d.[5]

[3] R. F. Christy, *Nucl. Phys.* **22,** 301 (1961).

[4] R. O. Bondelid and C. A. Kennedy, *Phys. Rev.* **115,** 1601 (1959).

[5] T. W. Bonner and C. F. Cook, *Phys. Rev.* **96,** 122 (1954).

Another procedure is to track slow neutron yield with de-excitation γ rays from the residual nucleus.[6]

Similar techniques have been used in other n-threshold studies, the interpretation of which is less apparent, for example,[7] in $Li^6(He^3, n)B^8$ and in $Li^7(\alpha, n)B^{10}$.

Review the modern state of the art in efficient and selective slow and fast neutron detectors and γ ray detectors. On this basis design experimental arrangements that might be suitable in reactions such as $Li^7(d, n)Be^8$ or $Be^9(d, n)B^{10}$, etc. On the basis of evidence available[8] attempt to determine which levels might exhibit strong threshold effects and which would be weak.

39.2. Partial Solutions

For S-wave production, the reaction intensity in the center-of-mass system is isotropic, thus:

$$I_c(\theta_c) = \alpha \qquad (\alpha \text{ a constant}).$$

Now $I_c(\theta_c)\, d\Omega_c = I(\theta)\, d\Omega$ (θ and θ_c are corresponding laboratory and center-of-mass angles). Hence,

$$I(\theta) = \alpha \frac{d\Omega_c}{d\Omega} = \alpha \frac{\sin \theta_c \, d\theta_c}{\sin \theta \, d\theta}.$$

It is straightforward to show that

$$\sin \theta \cos(\theta_c - \theta)\, d\theta_c = \sin \theta_c \, d\theta.$$

From which it follows that

$$I(\theta) = \frac{\alpha\{x \cos \theta \pm (1 - x^2 \sin^2 \theta)^{1/2}\}^2}{(1 - x^2 \sin^2 \theta)^{1/2}}$$

with $x = (v_{\text{c.m.}}/v_{3c}) = 1/\sin \theta_m$ (θ_m being the maximum laboratory emission angle near threshold). The $\pm$ refers to the double-valued nature of v_3 with θ. If the detector is assumed to have an efficiency per incident neutron of β/v_3 (β a constant), then weighting $I(\theta)$ gives, since

$$v_3 = v_{\text{c.m.}}\{\cos \theta \pm (1 - x^2 \sin^2 \theta)^{1/2}\}$$

[6] E. K. Warburton and L. F. Chase, *Nucl. Phys.* **34,** 517 (1962).

[7] K. L. Dunning, J. W. Butler, and R. O. Bondelid, *Phys. Rev.* **110,** 107 (1958).

[8] T. Lauritsen and F. Ajzenberg-Selove, *Nucl. Phys.* **78,** 1 (1966).

$$Y(\theta, \theta_m) = \frac{Kx^2 \cos \theta}{v_{\text{c.m.}} (1 - x^2 \sin^2 \theta)^{1/2}}; \qquad K = 2\alpha\beta$$

as the measured laboratory yield per unit solid angle.

While the relativistic treatment of the kinematic equations tend to be laborious for the general case, it is fairly straightforward to show:

$$E_{\text{TH}} = -Q \frac{(M_1 + M_2)}{M_2} (1 - \delta); \qquad \delta = \frac{Q}{2(M_1 + M_2)c^2}.$$

40. EXPERIMENTAL TARGETRY*

40.1. Problem

The differential energy loss of a heavy ion (i.e., mass greater than 1 amu) in penetrating matter can be written on a per atom basis as

$$\frac{dE}{dx} = -\varepsilon N \tag{40.1}$$

where N is the number of atoms per cubic centimeter and ε is the "differential atomic stopping cross section" (this designation of ε is historical; ε in fact is not the cross section of anything). Show that for the nonrelativistic case:

$$\varepsilon = 2.39 \times 10^{-16} \frac{z^2 Z_0 M}{E} \left[\ln \frac{2180E}{MI} - \frac{C_k(1/\eta)}{Z_0} \right] \text{eV-cm}^2 \tag{40.2}$$

where E, z, and M are the energy (in million electron volts), effective charge (in units of the electronic charge) and the mass (in atomic mass units) of the penetrating ion, respectively; Z_0, I, and $C_k(1/\eta)$ are the atomic charge, effective ionization potential (in electron volts), and a measure of the loss in contribution of the K electrons for the substance through which the ionizing particle is traveling.[1] The effective charge of the ion may be taken to correspond to full ionization if

$$\frac{E}{M} \gg 485 \left(\frac{z}{137}\right)^2 \text{MeV/amu.} \tag{40.3}$$

As a somewhat tighter criterion, when E/M is only 1.5 times the right-hand side the effective charge is only $\approx 0.6Z$ and when E/M is twice the right-hand side the effective charge is $\approx 0.9Z$, Z being the atomic charge of the penetrating ion. The $C_k(1/\eta)$ correction may be taken from Ref. 1 with $1/\eta = (470\, M/E)\, \{(Z_0 - 0.3)/137\}^2$ amu/MeV. In general, this correction $(470M/E)\ \{(Z_0 - 0.3)/137\}^2$ may be neglected if $1/\eta \lesssim 0.1$. Taking $I \approx 11.5\ Z_0$ eV is usually accurate enough.

[1] R. D. Evans, "The Atomic Nucleus," McGraw–Hill, New York, (1955).

* Problem 40 is by W. F. Hornyak.

When z cannot be taken for the fully ionized incident ion or when the $C_k(1/\eta)$ correction is substantial (i.e. $1/\eta \gtrsim 1$) it is better to use empirical curves (or semiempirical tables).[2] Many significant atomic effects and lattice effects (for single crystals) arise for slowly moving ions and may be involved in practical situations, for example, in thin "dE/dx" single crystal solid state detectors used in mass identification.[3]

For present purposes assume the expression for ε given is accurate enough for most target materials and incident ions. When the target consists of a molecular substance, there results:

$$\frac{1}{\rho}\frac{dE}{dx} = \frac{\Sigma_i(\nu_i\varepsilon_i)}{W} N_0 = \frac{\varepsilon_m N_0}{W} \text{ eV/g.} \tag{40.4}$$

Here $W = \Sigma_{ij}\nu_i\mu_{ij}M_{ij}$ is the molecular weight and ρ is the density (in grams per cubic centimeter), $N_0 = 6.03 \times 10^{23}$ (Avogadro's number), ν_i is the number of atoms of a particular type in the molecular structure, and μ_{ij} is the isotopic abundance of mass M_{ij} (atomic mass units). Clearly, ε_m is the differential molecular stopping cross section.

The basic determination of a cross section involves the measurement of the yield of particles (or γ rays) as a number per unit solid angle per incident particle for a target of thickness ΔX parallel to the beam direction. Then

$$\frac{\Delta Y}{\Delta\Omega}(\theta_c) = \frac{d\sigma}{d\Omega}(\theta_c)N_a\,\Delta X \tag{40.5}$$

where N_a is the number of nuclei of the proper type per cubic centimeter. Show that this may be rewritten as:

$$\frac{d\sigma}{d\Omega}(\theta_c, E) = \frac{1}{\mu_a\nu_a}\left(\frac{\varepsilon_m}{\Delta E}\right)\frac{\Delta Y}{\Delta\Omega}(\theta_c, E). \tag{40.6}$$

Here ΔE is the target thickness in electron volts, and it is assumed that $\Delta E \ll E$ and that ΔE is also small compared to any energy structure of interest in the cross section. Note that the quantity $(\varepsilon_m/\Delta E)$ is independent of energy E. Why?

Thus, if ΔE can be experimentally measured at any energy one need only compute ε_m at the same energy.

The yield from an infinitely thick target (neglecting scattering in the target) is just

[2] Studies in penetration of charged particles in matter, Publ. 1133. National Academy of Sciences (1964).

[3] H. W. Lewis, *Phys. Rev.* **125,** 937 (1962).

$$\frac{\Delta Y_\infty}{\Delta\Omega}(\theta_c, E_1) = \mu_a \nu_a \int_0^{E_1} \frac{d\sigma}{d\Omega}(\theta_c, E)\frac{dE}{\varepsilon_m(E)}. \tag{40.7}$$

Consider an experimental arrangement for measuring the total cross section for the reaction $Li^7(p, \alpha)He^4$ in which an evaporated (natural isotopic abundance) LiOH target on a thin nickel foil backing is used. The target is placed at 45° to the beam direction and the α particles are detected at 90° to the beam direction with a solid state detector subtending a half-angle of 5.0°. The number of α particles detected per 10^4 μC as a function of incident proton energy is shown in Table I. A thin foil was placed just

TABLE I.

E_p (keV)	Y_α (per 10^4 μC)	
25	2.2 ± 0.5	
50	188	Neglect error
75	1.48×10^3	
100	2.70×10^3	
150	1.64×10^4	
200	3.52×10^4	
250	5.96×10^4	
300	8.46×10^4	
350	1.12×10^5	
400	1.34×10^5	

in front of the solid state counter to stop all scattered protons from being counted while of course permitting the α particles to be detected. Calculate the highest energy scattered proton and the lowest energy α particle at $\theta = 90°$ and then determine a suitable foil for this purpose.

The same target set perpendicular to the beam direction was used in the $O^{16}(p, \gamma)F^{17}$ reaction to determine the target thickness. At the $E_p = 3.86$ MeV resonance the target thickness was determined to be 0.47 keV.

From this information calculate the total cross section in millibarns for the $Li^7(p, \alpha)He^4$ reaction as a function of energy. Assume an isotropic angular distribution.

If the beam current used had to be limited to 1 μA to prevent target deterioration, how many hours of running time was necessary to obtain the indicated statistical accuracy at $E_p = 25$ keV? In view of this calculation why was such a thin target used? To answer this question estimate the target thickness in kiloelectron volts at $E_p = 25$–100 keV (use empirical curves) and then consider the variation of cross section with energy.

Considering the fact that the exit channel consists of two identical Bose particles what is the lowest orbital angular momentum with which the reaction could proceed? Assuming that all vector-coupled angular momentum possibilities with this orbital angular momentum could occur, determine what terms in the following angular distribution expansion would be permitted (include interference possibilities)

$$Y(\theta, E) = Y(90^\circ, E)\,[1 + \sum_{n=1}^{\infty} A_n(E) \cos^n \theta]. \tag{40.8}$$

How do your conclusions change if the lowest two orbital angular momenta are considered? There is a prominent resonance in the $Li^7(p, \gamma)$ Be^8 reaction at $E_p = 441$ keV giving a 1^+ state in Be^{8*}. Would you expect to see any effect due to this level in the $Li^7(p, \alpha)He^4$ reaction? If it were determined empirically that $A_2(E) = +1.5 \times 10^{-3}E$ and $A_4(E) = -2.5 \times 10^{-4}E$, with E in kiloelectron volts and all other $A_n = 0$, recalculate the total cross section from the given yield at 90°.

Consider a thick target yield of γ rays from some reaction to be 10^{-5} γ rays per incident proton. What rate in equivalent curies does this represent when the beam current is 300 μA (1 C = 3.7×10^{10} γ rays/sec)?

Consider the case of looking at a single level Breit–Wigner resonance with a target of thickness Δ (Δ in energy units). Such a cross section would be

$$\sigma = \frac{\sigma_R \Gamma^2}{4(E - E_R)^2 + \Gamma^2} \tag{40.9}$$

and the yield per unit sphere would be

$$Y(E) = \mu_a \nu_a \sigma_R \Gamma \int_{E-\Delta}^{E} \frac{d\mathscr{E}}{\varepsilon_m(\mathscr{E})[4(\mathscr{E} - E_R)^2 + \Gamma^2]}. \tag{40.10}$$

If $\Delta \ll E_R$ then $\varepsilon_m(\mathscr{E}) \approx \varepsilon_m(E_R)$ a constant, and the integral may be readily evaluated. Calculate the energy E at which $Y(E)$ has its maximum value as a function of Δ. Determine the full width at half-maximum for $Y(E)$ as a function of Γ and Δ. Show that the maximum yield may be written as

$$Y_{max}(\Delta, \Gamma) = \frac{Y(\infty)}{\pi} \tan^{-1}\left\{\frac{2\Delta/\Gamma}{1 - (\Delta/\Gamma)^2}\right\} \tag{40.11}$$

and

$$Y(\Delta = \infty, \Gamma) = Y(\infty) = \frac{\pi \mu_a \nu_a \sigma_R \Gamma}{2\,\varepsilon_m(E_R)}. \tag{40.12}$$

Assume that $E_R = 1.000$ MeV and $\Gamma = 10$ keV, plot $\{Y(E)\,\varepsilon_m(E_R)\}/\{\mu_a v_a \sigma_R \Gamma\}$ as a function of E for $\Delta = 2, 10, 20, 50, \infty$ keV. Verify the general conclusions you derived above. In the case $\Delta = \infty$ what would be the result of taking $\varepsilon_m(\mathscr{E}) \approx K/\mathscr{E}$?

Consider how you could use the above calculations in determining Δ if $\Delta \gtrsim \Gamma$ and also if $\Delta \lesssim \Gamma$. What effect would you expect the discreteness of the energy loss processes to have on the above calculations?[3]

41. DYNAMICS OF MASS EXCHANGE REACTIONS*

41.1. Problem

Direct process model calculations involving an exchange of a mass fragment between incident projectile and target nucleus in the general case are complicated by the relative importance of various effective interaction potentials.[1] As an example of the importance of kinematic factors in determining observable angular distributions we will examine a coherent pair of processes referred to as the "direct stripping" process and the "heavy particle stripping" process.[2, 3] The general nuclear reaction may be considered as $M_2(M_1, M_3)M_4$ with M_1 the incident particle with laboratory energy E_1, M_2 the target, and usually M_3 a light fragment and M_4 a heavy fragment both in the exit channel. Direct stripping imagines M_3 to have been stripped from M_1, the remaining fragment $(M_1 - M_3)$ being simultaneously captured by M_2 to form M_4. In heavy particle stripping M_3 is stripped from M_2 by M_1 which in turn is captured by the remaining core $(M_2 - M_3)$. These processes are coherent and interference effects will also arise.

As a starting point, one should review the quantum mechanics of an N-particle system with the interaction potential depending only on the relative position of the particles. For such a system of N particles $(N - 1)$ relative coordinates and the center-of-mass coordinate may be used to achieve the separation of the Schrödinger equation.[4] Consider two particles M_1, $\mathbf{R}_1$, and $\mathbf{P}_1$ (mass, coordinate, and momentum) and M_2, $\mathbf{R}_2$, and $\mathbf{P}_2$. If one defines

$$\mathbf{R} = \frac{M_1\mathbf{R}_1 + M_2\mathbf{R}_2}{M_1 + M_2}, \qquad \mathbf{r} = \mathbf{R}_1 - \mathbf{R}_2 \tag{41.1a}$$

[1] F. Ajzenberg-Selove, ed., "Nuclear Spectroscopy" Part B, Chapter V.B.2. Academic Press, New York, 1960.

[2] L. Madansky, and G. Owen, *Phys. Rev.* **99,** 1608 (1955).

[3] G. Owen, and L. Madansky, *Phys. Rev.* **105,** 1766 (1957).

[4] A. Messiah, "Quantum Mechanics," Vol. I, Chapter IX, (11–13). North-Holland Publ., Amsterdam, 1964.

* Problem 41 is by W. F. Hornyak.

$$M = M_1 + M_2, \qquad \mu = \frac{M_1 M_2}{M_1 + M_2},$$
$$\mathbf{P} = \mathbf{P}_1 + \mathbf{P}_2, \qquad \mathbf{p} = \frac{M_2\mathbf{P}_1 - M_1\mathbf{P}_2}{M_1 + M_2}, \tag{41.1b}$$

show that

$$M_1 M_2 = \mu M$$
$$\frac{P^2}{2M} + \frac{p^2}{2\mu} = \frac{P_1^{\,2}}{2M_1} + \frac{P_2^{\,2}}{2M_2}$$
$$\mu r^2 + MR^2 = M_1 R_1^{\,2} + M_2 R_2^{\,2} \tag{41.2}$$
$$\mathbf{p}\cdot\mathbf{r} + \mathbf{P}\cdot\mathbf{R} = \mathbf{P}_1\cdot\mathbf{R}_1 + \mathbf{P}_2\cdot\mathbf{R}_2$$
$$\mathbf{R}\times\mathbf{P} + \mathbf{r}\times\mathbf{p} = \mathbf{R}_1\times\mathbf{P}_1 + \mathbf{R}_2\times\mathbf{P}_2.$$

In the reaction $M_2(M_1, M_3)M_4$ with incident laboratory energy E_1, show

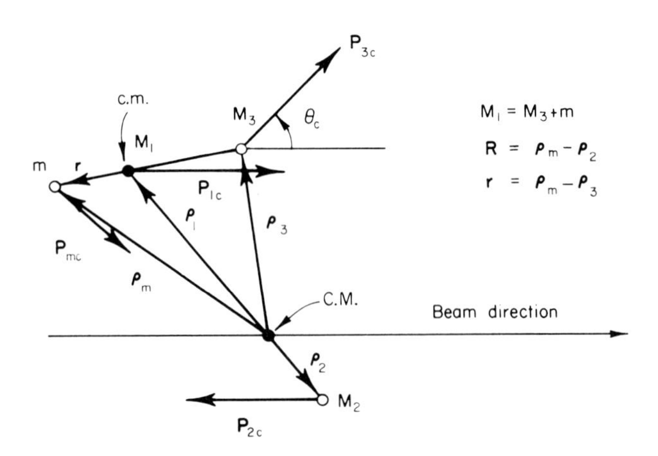

FIG. 41.1. Direct stripping process.

that in the center-of-mass system

$$p_{1c}^{\,2} = \left(\frac{M_2\mathbf{P}_1 + M_1\mathbf{P}_2}{M_1 + M_2}\right)^2 = M_1^{\,2}v_{1c}^{\,2} = M_2^{\,2}v_{2c}^{\,2} = p_{2c}^{\,2}$$
$$= 2\frac{M_1 M_2}{(M_1 + M_2)}\frac{M_2 E_1}{(M_1 + M_2)} = 2\mu_{12}E \tag{41.3}$$

where E is the kinetic energy in the center-of-mass. Also

$$p_{3c}{}^2 = M_3{}^2 v_{3c}{}^2 = M_4{}^2 v_{4c}{}^2 = p_{4c}{}^2 = 2\,\frac{M_3 M_4}{(M_3 + M_4)}\left[\frac{M_2 E_1}{M_1 + M_2} + Q\right]$$
$$= 2\mu_{34}[E + Q] \qquad (41.4)$$

where as usual $Q = (\Delta M)c^2$.

Consider the two interacting particles M_1 and M_2 in the center-of-mass system with momenta $\mathbf{p}_{1c}$ and $\mathbf{p}_{2c}$ found above (note $\mathbf{p}_{1c} = -\mathbf{p}_{2c}$). Let m be the mass of the fragment transferred from M_1 to M_2; thus, $M_1 = M_3 + m$ and $M_4 = M_2 + m$. Then referring to Fig. 41.1 and taking the center-of-mass as the origin and considering the relative coordinates $\mathbf{r} = \boldsymbol{\rho}_m - \boldsymbol{\rho}_3$ and $\mathbf{R} = \boldsymbol{\rho}_m - \boldsymbol{\rho}_2$ there results for the relative coordinate $\mathbf{r}$ a physical system with reduced mass $mM_3/(m + M_3)$ and relative momentum $\hbar\pi_d = (M_3/M_1)\,\mathbf{p}_{1c} - \mathbf{p}_{3c}$ and in the relative coordinate $\mathbf{R}$ a physical system with reduced mass $mM_2/(m + M_2)$ and relative momentum (referred to as the momentum transfer) $\hbar\mathbf{q}_d = \mathbf{p}_{1c} - (M_2/M_4)\mathbf{p}_{3c}$. Show that these relations are so. This is the case of direct stripping. In plane

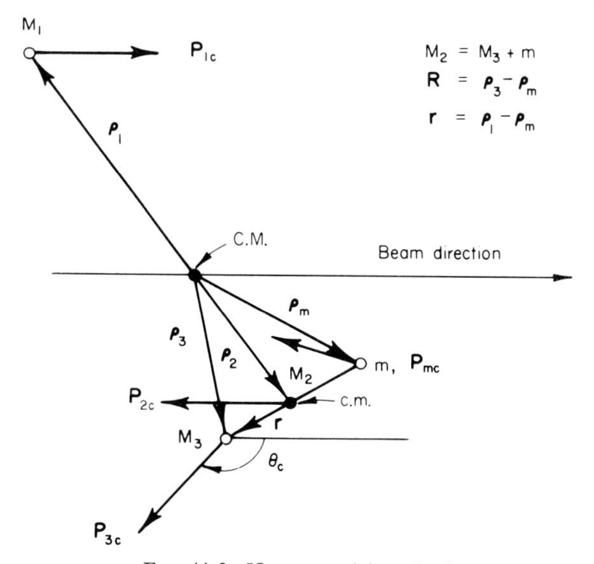

FIG. 41.2. Heavy particle stripping.

wave Born approximation the first system (the internal M_1 problem) largely determines the magnitude of the cross section while the second largely determines the angular distribution through a centrifugal barrier factor $j_L(q_d R_d)$, L being the appropriate orbital angular momentum and R_d the channel radius.

For heavy particle stripping refer to Fig. 41.2. Here we write $M_2 = M_3 + M$ and $M_4 = M_1 + M$ and take as relative coordinates $\mathbf{r} = \boldsymbol{\rho}_3 - \boldsymbol{\rho}_m$ and $\mathbf{R} = \boldsymbol{\rho}_3 - \boldsymbol{\rho}_m$.

Show that the relative systems have reduced mass $mM_3/(m + M_3)$ and relative momentum $\hbar\boldsymbol{\pi}_h = \mathbf{p}_{3c} + (M_3/M_2)\mathbf{p}_{1c}$ for one, and for the other reduced mass $mM_1/(m + M_1)$ and relative momentum (momentum transfer) $\hbar\mathbf{q}_h = \mathbf{p}_{1c} + (M_1/M_4)\mathbf{p}_{3c}$. Neglecting internal motion factors the differential cross section then becomes

$$\frac{d\sigma}{d\Omega}(\theta_c) = \text{const} \times \left| j_L(q_d R_d) + \alpha j_L'(q_h R_h) \right|^2. \qquad (41.5)$$

The order of the spherical Bessel functions L and L' is determined by selection rules involving the spins and parities of the interacting particles and fragments. What are these selection rules?

For orientation we will apply this simple approach to the $B^{11}(d, n)C^{12}gd$ reaction at $E_1 = 5.000$ MeV. Show that the wave number in a frame of reference in which a "particle" has energy E million electron volts and reduced mass μ (atomic mass units) is

$$k = 0.220(E\mu)^{1/2} \qquad \text{in units of } F^{-1}.$$

Show that for the direct stripping process $L = 1$ is uniquely determined. Show that for heavy particle stripping $L' = 0$ can only result when $B^{11}gd$ is taken to have a $B^{10*} + n$ parentage with $J^\pi = 1^+$ for B^{10*} (for example, the first excited state) coupled to a $p_{3/2}$ neutron. What would be the lowest value of L' if the parentage were $B^{10}gd + n$?

Calculate the contributions of direct and heavy particle stripping (and the interference term) if $R_d = R_h = 4.50$ F, $L = 1$, $L' = 0$, and $\alpha + 1.60$ real. Take

$$\frac{d\sigma}{d\Omega}(\theta_c) = 100 \left| j_L(q_d R_d) + \alpha j_L'(q_h R_h) \right|^2. \qquad (41.6)$$

41.2. Partial Solution

Taking $E_1 = 5.000$ MeV, $Q_0 = 13.731$ MeV, $L_d = 1$, $L_h = 0$, $R_d = 4.50$ F, there results:

$$(q_d R_d)^2 = 20.63 - 19.58 \cos\theta_c$$

$$(q_h R_h)^2 = 7.47 + 3.55 \cos\theta_c$$

and

$$\frac{d\sigma}{d\Omega}(\theta_c) = 100 \, | \, j_1(q_d R_d) + 1.60 \, j_0(q_h R_h) \, |^2 .$$

The calculated values of $(d\sigma/d\Omega)(\theta_c)$ are given in Fig. 41.3. Standard tables for the spherical Bessel functions are available.[5]

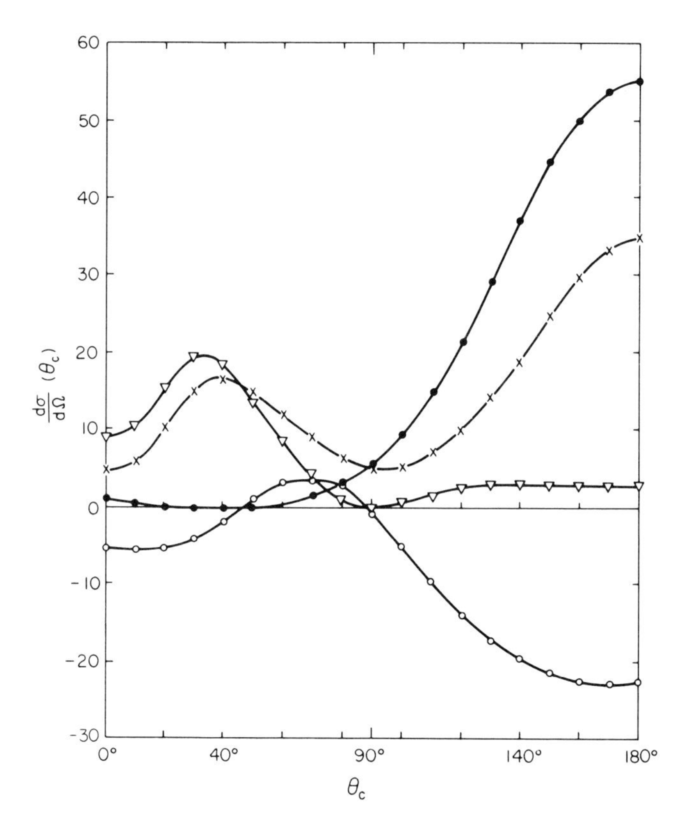

FIG. 41.3. Direct and heavy particle stripping amplitudes: ▽ — ▽ represents direct intensity, ● — ● heavy particle intensity, ○ — ○ interference intensity, and × — × total intensity.

[5] M. Abromovitz, and I. Stegun, ed., "Handbook of Mathematical Functions." Dover, New York, 1964.